Technology and Social Inclusion

Technology and Social Inclusion
Rethinking the Digital Divide

Mark Warschauer

The MIT Press
Cambridge, Massachusetts
London, England

This book was set in Sabon by SNP Best-set Typesetter Ltd., Hong Kong Printed and bound in the United States of America.

Library of Congress Cataloging-in-Publication Data

Warschauer, Mark.
 Technology and social inclusion : rethinking the digital divide /
Mark Warschauer.
   p. cm.
 Includes bibliographical references and index.
 ISBN 0-262-23224-3 (hc. : alk. paper)
 1. Digital divide. 2. Marginality, Social. I. Title.
HN49.I56 W37 2003
303.48′33—dc21   2002075130

10 9 8 7 6 5 4 3 2 1

For Keiko

Contents

Acknowledgments

This book has been nearly ten years in the making, and its completion would not have been possible without the great deal of institutional and personal support I received over those years.

Most of the actual writing took place in 2001, while I worked as an assistant professor in the Department of Education at the University of California, Irvine. The department has been quite generous in allowing me to devote time and resources to this book, and I would especially like to thank the chair of the department, Rudy Torres; the former chair, Louis Mirón; and the former acting chair, Robert Beck, for their full support. Other colleagues in the department have also been extremely helpful, including Hank Becker, Joan Bissell, and Ann De Vaney.

I am also affiliated with the UC Irvine Center for Research on Information Technology and Organizations (CRITO), and my participation in CRITO seminars has been very helpful to the development of my ideas. I would like to thank CRITO and its former acting director, James Danziger, for involving me in the center and its seminars.

During the summer of 2001, I took research trips to China, India, and Brazil. I would like to thank all the individuals who took time to meet with me, and especially to thank the following people who helped with my arrangements and provided invaluable advice and assistance: Dr. Chen Hong of Beijing Capital Normal University in China; Chetan Sharma of Datamation in New Delhi; Naveen Prakash of Gyandoot Samiti in Dhar, India; Senthil Kumaran of M. S. Swaminathan Foundation in Madras, India; Aditya Dev Sood of the Center for Knowledge Societies in Bangalore, India; Vera Mello of the University of São Paulo, Brazil; and Solange Gervai of Yázigi Internexus in São Paulo.

During 2001–2002, I had four grants from the University of California to examine the availability of, access to, and use of information and communication technologies in low-income neighborhood schools. This book has benefited from that research, and I would like to acknowledge the support of UC Nexus, directed by Charles Underwood of the University of California's President's Office; the University of California All Campus Consortium on Research on Diversity, directed by Jeannie Oakes and Danny Solorzano; and the University of California, Irvine, Center for Educational Partnerships, directed by Juan Lara. I would also like to acknowledge and thank LeeAnn Stone, Michele Knobel, Melanie Wade, Fang Xu, and Jodie Wales for their collaboration on that research.

In 2001, I also visited several Community Technology Centers in California. Linda Fowells and Richard Chabran of Community Partners provided extremely helpful support and information for this aspect of my research.

From 1998 to 2001, I worked for the America-Mideast Educational and Training Services, Inc. (AMIDEAST) on a project in Egypt called IELP-II, funded by the U.S. Agency for International Development (USAID). My work and research in Egypt was central to many of the ideas expressed in this book, and I would like to thank AMIDEAST, USAID, and IELP-II for the opportunity to carry out that work. I would especially like to thank Doug Duncan of AMIDEAST Washington, Jennifer Notkin of USAID Egypt, and Randa Effat of IELP-II. Ghada Refaat and Ayman Zohry of IELP-II assisted me in carrying out a research study of online language use in Egypt, and Russanne Hozayin of IELP-II also was generous with her information and assistance. I would also like to thank a number of other Egyptian and American colleagues in Egypt who collaborated with me and otherwise provided information and support, including Richard Boyum of the U.S. Embassy; Inas Barsoum of Ain Shams University; Kamal Fouly of Minia University; Maha El Said and Mounira Soliman of Cairo University; and Paul Stevens, Gini Stevens, and Kate Coffield of the American University in Cairo.

From 1994 to 1998, I taught and conducted research at the University of Hawai'i for the Department of Second Language Studies, the

National Foreign Language Resource Center (NFLRC), and the College of Languages, Linguistics, and Literature. Again, much of that research has influenced, and found its way into, this book, and I would like to thank Roderick Jacobs, dean of the College; Richard Schmidt, director of the NFLRC; and Gabrielle Kasper, Graham Crookes, and Kathryn Davis of the Department of Second Language Studies for their support. Lois Yamauchi of the University of Hawai'i Educational Psychology Department and Jim Cummins of the University of Toronto also provided valuable support for my research in Hawai'i. I received a great deal of assistance from colleagues in the Hawaiian educational community for my research, especially from Keola Donaghy of Hale Kuamo'o at the University of Hawai'i, Hilo, and Makalapua Ka'awa of the University of Hawai'i, Manoa.

A number of colleagues provided assistance, support, and ideas by co-authoring pieces that formed the basis of this book, critiquing this or related manuscripts, or otherwise sharing ideas. Rick Kern, Hank Becker, Phil Agre, Dafney Dabach, Eszter Hargittai, Fang Xu, and Martha Forero offered extremely helpful comments on the manuscript. Their contribution has been enormous, and I cannot thank them enough. I would also like to recognize and thank Dorothy Chun of the University of California, Santa Barbara; Mike Cole, Olga Vasquez, Bud Mehan, and Leigh Starr of the University of California, San Diego; Irene Thompson and Pam DaGrossa of *Language Learning and Technology* journal; Peiya Gu of Suzhou University in China; and Heidi Helfand of Expert City for their collaboration and support in a variety of projects and discussions that helped give intellectual birth to this book. And I am indebted to Cathy Appel for her invaluable assistance with the manuscript preparation.

The editors and reviewers at MIT Press have been extremely helpful, and I would especially like to thank acquisition editor Katherine Innis. MIT Press has shown great leadership in areas related to the social, economic, political, and cultural ramifications of new technology use, and the editors and associates at the Press have been remarkably helpful in ushering this book from original proposal to publication.

My deepest thanks are to my wife, Keiko Hirata. Keiko has read and provided extremely helpful comments on almost everything I have written, and her own research on civil society in East Asia and the Middle East has played a formative role in my work. I could not have written this book without her, nor would such an endeavor have been worthwhile without her love and support.

Technology and Social Inclusion

Introduction

The purpose of this book is to examine the relationship between information and communication technology (ICT) and social inclusion. A starting point for my research has been the concept of a digital divide, used by the U.S. National Telecommunications and Information Administration under the Clinton administration to refer to the gap between those who do and do not have access to computers and the Internet. However, during the process of my research, the notion of a digital divide and its logical implication—that social problems can be addressed through providing computers and Internet accounts—have seemed increasingly problematic. Three vignettes will help illustrate this point.

A Slum "Hole-in-the-Wall"

In 2000 the government of New Delhi, in collaboration with an information technology corporation, established a project, known as the Hole-in-the-Wall experiment, to provide computer access to the city's street children.[1] An outdoor five-station computer kiosk was set up in one of the poorest slums of New Delhi. Though the computers themselves were inside a booth, the monitors protruded through holes in the walls, as did specially designed joysticks and buttons that substituted for the computer mouse. Keyboards were not provided. The computers were connected to the Internet through dial-up access. A volunteer inside the booth helped keep the computers and Internet connections running.

No teachers or instructors were provided, in line with a concept called minimally invasive education. The idea was to allow the children unfettered 24-hour access to learn at their own pace and speed rather than tie them to the directives of adult organizers or instructors.

According to reports, children who flocked to the site taught themselves basic computer operations. They worked out how to click and drag objects; select different menus; cut, copy, and paste; launch and use programs such as Microsoft Word and Paint; get on the Internet; and change the background "wallpaper." The program was hailed by researchers (e.g., Mitra 1999) and government officials[2] as a groundbreaking project that offered a model for how to bring India's and the world's urban poor into the computer age.

However, visits to the computer kiosk indicated a somewhat different reality. The Internet access was of little use because it seldom functioned. No special educational programs had been made available, and no special content was provided in Hindi, the only language the children knew. Children did learn to manipulate the joysticks and buttons, but almost all their time was spent drawing with paint programs or playing computer games.

There was no organized involvement of any community organization in helping to run the kiosk because such involvement was neither solicited nor welcomed (see chapter 6). Indeed, the very architecture of the kiosk—based on a wall rather than in a room—made supervision, instruction, and collaboration difficult.

Parents in the neighborhood had ambivalent feelings about the kiosk. Some saw it as a welcome initiative, but most expressed concern that the lack of organized instruction took away from its value. Some parents even complained that the kiosk was harmful to their children. As one parent stated, "My son used to be doing very well in school, he used to concentrate on his homework, but now he spends all his free time playing computer games at the kiosk and his schoolwork is suffering." In short, parents and the community came to realize that minimally invasive education was, in practice, minimally effective education.

An Information Age Town

In 1997, Ireland's national telecommunications company held a national competition to select and fund an "Information Age Town."[3] A major rationale behind the effort was to help overcome the gap between Ireland's emerging status as a multinational business center of ICT *production* and the rather limited *use* of ICT among Ireland's own people and indigenous small businesses.

Towns of 5,000 people and more across Ireland were invited to compete by submitting proposals detailing their vision of what an Information Age Town should be and how they could become one. The winning town was to receive 15 million Irish pounds (at that time roughly $22 million U.S. dollars—USD) to implement its vision.

The sponsor of the competition, Telecom Eirann (later renamed Eircom), was getting ready to be privatized. The company naturally had an interest in selecting the boldest, most ambitious proposal so as to showcase the winning town as an innovative example of what advanced telecommunications could accomplish for the country under the company's leadership. Four towns were chosen as finalists, and then Ennis, a small, remote town of 15,000 people in western Ireland, was selected among them as the winner. The prize money that Ennis received represented over $1,200 USD per resident, a huge sum for a struggling Irish town.

At the heart of Ennis's winning proposal was a plan to give an Internet-ready personal computer to every family in the town. Other initiatives included an ISDN line to every business, a Web site for every business that wanted one, smart-card readers for every business (for a cashless society), and smart cards for every family. Ennis was strongly encouraged by Telecom Eivann to implement these plans as quickly as possible.

Meanwhile, the three runners-up—the towns of Castlebar, Kilkenny, and Killarney—each received consolation prizes of 1 million Irish pounds (about $1.5 million USD). These towns were given as much time as they needed to make use of the money.

How did the project turn out? A visit to Ennis three years later by a university researcher indicated that the town had little to show for its money. Advanced technology had been thrust into people's hands with little preparation. Training programs had been run, but they were not sufficiently accompanied by awareness programs as to why people should use the new technology in the first place. And, in some instances, well-functioning social systems were disrupted in order to make way for the showcase technology.

For example, as is the case in the rest of Ireland, the unemployed of Ennis had been reporting to the social welfare office three times a week to sign in and receive payments. Following their visits, the people usually stayed around the office to chat with other unemployed workers. The sign-in system thus facilitated an important social function to overcome the isolation of the unemployed.

As part of the "Information Age Town" plan, though, the unemployed received computers and Internet connections at home. They were instructed to sign in and receive electronic payments via the Internet rather than come to the office to sign in. But many of the unemployed couldn't figure out how to operate the equipment, and most others saw no reason to do so when it deprived them of an important opportunity for socializing. A good number of those computers were reportedly sold on the black market, and the unemployed simply returned of their own accord to coming to the social welfare office to sign in.

Meanwhile, what happened in the other three towns? With far fewer resources, they were forced to carefully plan how to make use of their funds rather than splurging for massive amounts of equipment. Community groups, small businesses, and labor unions were involved in the planning process. Much greater effort and money were spent on developing awareness, planning and implementing effective training, and setting up processes for sustainable change rather than merely on purchase of equipment. The towns built on already existing networks among workers, educators, and businesspeople to support grassroots uses of technology for social and economic development. Information about social services and job opportunities was put online. Small businesses and craft workers learned how to pool their resources to promote their products through e-commerce. Technology coordinators were appointed at schools and worked with other teachers to develop plans for better integration of ICT in classrooms. In the end, according to a researcher from University College Dublin,[4] the three runners-up, which each received only one-fifteenth of the money that Ennis received, actually had more to show for their efforts to promote social inclusion through technology than did the winner.

A Model Computer Lab

An international donor project funded by the U.S. Agency for International Development (USAID) decided to donate a computer laboratory to the college of education at a major Egyptian university. The purpose of the donation was to establish a model teacher training program in computer-assisted learning in one of the departments of the college. State-of-the-art equipment was selected, including more than forty Pentium III computers, an expensive video projection system, several printers and scanners, and tens of thousands of U.S. dollars worth of educational software. This was to be a model project that both the U.S. and Egyptian governments would view with pride. To guarantee that the project would be sustainable, the Egyptian university would be required to manage all the ongoing expenses and operations, including paying for Internet access, maintaining the local area network (LAN), and operating the computer laboratory.

Under a paid contract from USAID, a committee from the college of education put together a detailed proposal on how the laboratory would be used, run, and maintained. Based on this proposal, USAID purchased all the hardware and software. However, well before the equipment was installed, it became clear that the college would have difficulty absorbing such a huge and expensive donation. Other departments within the college, which together had access to only a handful of computers, became envious that a single department would have such modern and expensive equipment, and they attempted to block the university's support for the lab. The college of education and the university could not easily justify spending the money to house and maintain such an expensive laboratory for a single program when other programs were poorly funded. No money was available to hire an outside LAN manager or provide Internet access at the level agreed upon in the proposal. Faculty relations problems also arose, as a key department chair resented the involvement and initiative of less senior faculty members who were taking computer training and working together to plan new curricula. Because of all these difficulties, the expensive state-of-the-art computers sat in boxes in a locked room for more than a year before they were even installed, thus losing about one-third of their economic value.

Rethinking the Digital Divide

Each of the programs described in the preceding vignettes was motivated by a sincere attempt to improve people's lives through ICT. But each program ran into unexpected difficulties that hindered the results. Of course, any ICT project is complicated, and none can be expected to run smoothly. But the problems with these projects were neither isolated nor random. Rather, these same types of problems occur again and again in technology projects around the world, which too often focus on providing hardware and software and pay insufficient attention to the human and social systems that must also change for technology to make a difference. As seen in these three vignettes, meaningful access to ICT comprises far more than merely providing computers and Internet connections. Rather, access to ICT is embedded in a complex array of factors encompassing physical, digital, human, and social resources and relationships. Content and language, literacy and education, and community and institutional structures must all be taken into account if meaningful access to new technologies is to be provided.

Some would try, as I have tried in the past, to stretch the notion of a digital divide to encompass this broad array of factors and resources. In this sense, a digital divide is marked not only by physical access to computers and connectivity but also by access to the additional resources that allow people to use technology well. However, the original sense of *digital divide*, which attached overriding importance to the physical availability of computers and connectivity rather than to issues of content, language, education, literacy, or community and social resources, is difficult to overcome in people's minds.

A second problem with the digital divide concept is its implication of a bipolar societal split. As Cisler (2000) argues, there is not a binary division between information haves and have-nots, but rather a gradation based on different degrees of access to information technology. Compare, for example, a professor at UCLA with a high-speed connection in her office, a student in Seoul who occasionally uses a cyber café, and a rural activist in Indonesia who has no computer or phone line but whose colleagues in the nongovernmental organization (NGO) with whom she is working download and print out information for her. This example illustrates just three degrees of possible access a person can have to online material.

The notion of a binary divide between haves and have-nots is thus inaccurate and can even be patronizing because it fails to value the social resources that diverse groups bring to the table. For example, in the United States, African Americans are often portrayed as being on the wrong side of a digital divide (e.g., Walton 1999) when in fact Internet access among blacks and other minorities varies tremendously by income group—with divisions between blacks and whites decreasing as income increases (NTIA 2000). Some argue that the stereotype of disconnected minority groups could even serve to further social stratification by discouraging employers or content providers from reaching out to those groups. As Henry Jenkins, director of comparative media studies at the Massachusetts Institute of Technology, argues, "The rhetoric of the digital divide holds open this division between civilized tool-users and uncivilized nonusers. As well meaning as it is as a policy initiative, it can be marginalizing and patronizing in its own terms" (quoted in Young 2001, A51).

In addition, the notion of a digital divide—even in its broadest sense— implies a chain of causality: that lack of access (however defined) to computers and the Internet harms life chances. While this point is undoubtedly true, the reverse is equally true; those who are already marginalized will have fewer opportunities to access and use computers and the Internet. In fact, technology and society are intertwined and co-constitutive, and this complex interrelationship makes any assumption of causality problematic.

Finally, the digital divide framework provides a poor road map for using technology to promote social development because it over-emphasizes the importance of the physical presence of computers and connectivity to the exclusion of other factors that allow people to use ICT for meaningful ends. Rob Kling, director of the Center for Social Informatics at Indiana University, explains this shortcoming well:[5]

[The] big problem with "the digital divide" framing is that it tends to connote "digital solutions," i.e., computers and telecommunications, without engaging the important set of complementary resources and complex interventions to support social inclusion, of which informational technology applications may be enabling elements, but are certainly insufficient when simply added to the status quo mix of resources and relationships.

The bottom line is that there is no binary divide and no single over-riding factor for determining such a divide. ICT does not exist as an external variable to be injected from the outside to bring about certain results. Rather, it is woven in a complex manner into social systems and processes. And, from a policy standpoint, the goal of using ICT with marginalized groups is not to overcome a digital divide but rather to further a process of social inclusion. To accomplish this, it is necessary to "focus on the transformation, not the technology" (Jarboe 2001, 31). For all these reasons, I join with others (e.g., DiMaggio and Hargittai 2001; Jarboe 2001) in recognizing the historical value of the digital divide concept (it helped focus attention on an important social issue) while preferring to embrace alternative concepts and terminology that more accurately portray the issues at stake and the social challenges ahead.

Social Inclusion

The alternative framework that I suggest in this book is the intersection of ICT and social inclusion. Social inclusion and exclusion are promi-nent concepts in European discourse.[6] They refer to the extent that indi-viduals, families, and communities are able to fully participate in society and control their own destinies, taking into account a variety of factors related to economic resources, employment, health, education, housing, recreation, culture, and civic engagement.

Social inclusion is a matter not only of an adequate share of resources but also of "participation in the determination of both individual and collective life chances" (Stewart 2000). It overlaps with the concept of socioeconomic equality but is not equivalent to it. There are many ways that the poor can have fuller participation and inclusion even if they lack an equal share of resources. At the same time, even the well-to-do may face problems of social exclusion because of political persecution or dis-crimination based on age, gender, sexual preference, or disability. The concept of social inclusion does not ignore the role of class but recog-nizes that a broad array of other variables help shape how class forces interact. Though a historical treatment of the term is beyond the scope of this book, one could argue that the concept of social inclusion reflects particularly well the imperatives of the current information era, in which issues of identity, language, social participation, community, and civil society have come to the fore (Castells 1997).

This book takes as a central premise that the ability to access, adapt, and create new knowledge using new information and communication technology is critical to social inclusion in today's era (see chapter 1). I thus examine several questions related to this premise: How and why is access to ICT important for social inclusion? What does it mean to have access to ICT? How can access for social inclusion best be promoted in diverse circumstances? By focusing on technology for social inclusion, I thus hope to help reorient discussion of the digital divide from one that focuses on gaps to be overcome by provision of equipment to one that focuses on social development issues to be addressed through the effective integration of ICT into communities, institutions, and societies.

Sources of Data

This book draws largely on my own empirical research in a number of countries throughout the world. I have focused most of my research on countries such as India, Brazil, Egypt, China, and the United States that have extensive poverty; large gaps between rich and poor; substantial but unequally distributed ICT resources; and a myriad of local and national programs attempting to use technology to promote social inclusion.

My empirical research in these countries has included both long-term ethnographic research (e.g., in Hawai'i, 1995–1997 and Egypt, 1998–2001) and short-term, intensive field observations (e.g., in India, 2001; Brazil, 2001; China, 1999 and 2001; and California, 2001). During this widespread and differential research, I visited schools, universities, community technology centers, telecenters, NGOs, and government offices. I observed dozens of technology access and training programs. I interviewed a wide range of people, including government officials and policymakers, educators, representatives of community associations and NGOs, leaders of information technology companies, and children and adults participating in community technology programs. Altogether, over a period of six years I interviewed more than 200 people and wrote up some 500 pages of observation field notes.

In addition, I bring to this extensive data set an analysis of secondary data published in a variety of print and online sources. These include newspaper and magazine articles, books, academic journals, governmental and nongovernmental reports, and online essays and discussions. Some of my most important ongoing sources have included periodicals such as the *New York Times*, the *Los Angeles Times*, and the *Economist*; online forums such as the Digital Divide discussion list, the Global Knowledge for Development discussion list, the Association for Internet Researchers discussion list, and Red Rocker Eater News;[7] and reports from the World Bank, the United Nations Development Programme, and the National Telecommunications and Information Administration.

Organization

The first two chapters of this book provide a historical and theoretical framework to issues of technology and social inclusion. Chapter 1 provides the overall contextual background for the book by analyzing the transformation occurring in global economics, society, and technology. Chapter 2 looks back to other historical divides, such as those related to electrification, universal telephone service, and literacy, to analyze models of access to technology and media. This second chapter identifies four types of technology-associated resources that are essential to access and inclusion: physical, digital, human, and social. Subsequently, chapters 3 to 6 analyze these four resources in more depth. Chapter 3 examines physical resources: computers and connectivity. Chapter 4 examines digital resources: content and language. Chapters 5 examines human resources: literacy and education; and chapter 6 examines social resources: communities and institutions. Each of these chapters attempts to provide both a conceptual framework, drawing on relevant social theory, and empirical evidence and examples from both developing and developed countries. Finally, chapter 7 draws together the main arguments of the book by examining theories of the social embeddedness of technology.

1

Economy, Society, and Technology: Analyzing the Shifting Terrains

The concept of a digital divide gained headway in the mid-to-late 1990s, at the same time that the Internet and dot-com booms were under way in the United States. In a sense, the digital divide approach—which often emphasized getting people connected anyway they could at all cost so that they wouldn't be left behind—reflected the general spirit of the times, which were based on a superficial understanding of the Internet's relationship to economic and social change. At an economic level, too much emphasis was being put on the so-called Internet economy, reflected in the wild surge of dot-com businesses, many of which went bankrupt after failing to earn a single dollar. At the societal level, the hottest idea was that of cyberspace, supposedly an entirely different plane of existence (e.g., Barlow 1996). Both of these perspectives reflected the errant view that information and communication technology (ICT) was creating a parallel reality and that it was thus necessary for people to make the leap across the divide from the old reality to the new one in order to succeed.

On the tail end of the dot-com boom, with the NASDAQ (technology stock exchange) having fallen more than 60% from its high point, and computer sales and Internet growth rates leveling off in the United States, it is easy to dismiss the entire era as a passing fad that should be put behind us. From this view, the digital divide can be seen as either passé (because most people who want computer and Internet access in the richest countries can now afford them) or irrelevant (because those who don't have Internet access don't really need it). A related perspective is to portray computers and the Internet as mere devices, without any particular public import—a perspective that led one

Bush administration official to compare the digital divide to a "Mercedes divide."[1]

While this cynicism may reflect a corrective to the misguided Internet obsession of the late 1990s, it too is mistaken. Although ICT has not created a parallel world that one must leap into at all cost, it has contributed to a profound change in the real world we live in. While the dot-com economy has gone bust, the underlying information economy surges on. While notions of cyberspace fade away, real-life applications of e-commerce, e-governance, and Internet-enhanced learning thrive. And while the current U.S. administration does not emphasize a digital divide, many governments around the world are stressing the importance of ICT for social inclusion.[2]

The shift from a focus on a digital divide to social inclusion rests on three main premises: (1) that a new information economy and network society have emerged; (2) that ICT plays a critical role in all aspects of this new economy and society; and (3) that access to ICT, broadly defined, can help determine the difference between marginalization and inclusion in this new socioeconomic era.

Informationalism

What is new in the U.S. and world economies is not just a rise and fall of dot-com businesses but rather a deeper and more long-lasting transformation: the emergence of a new stage of global capitalism. This new stage, referred to by some as postindustrialism (e.g., Bell 1973), has been labeled informationalism by Castells (2000b).[3] Informationalism represents a third industrial revolution (table 1.1). The first followed the invention of the steam engine in the eighteenth century and was characterized by the replacement of hand tools by machines, mostly in small workshops (Singer et al. 1958, cited in Castells 2000b). The second followed the harnessing of electricity in the nineteenth century and was characterized by the development of large-scale factory production (Mokyr 1990, cited in Castells 2000b). The third revolution came to fruition in the 1970s with the diffusion of the transistor, the personal computer, and telecommunications. In other words, what we have is not

Table 1.1
The Three Industrial Revolutions

	First Industrial Revolution	Second Industrial Revolution	Third Industrial Revolution
Beginning	Late 18th century	Late 19th century	Mid-to-late 20th century
Key technologies	Printing press, steam engine, machinery	Electricity, internal combustion, telegraph, telephone	Transistor, personal computers, telecommunications, Internet
Archetypical workplace	Workshop	Factory	Office
Organization	Master-apprentice-serf	Large vertical hierarchies	Horizontal networks

an Internet economy but an information economy in which computers and the Internet play an essential enabling role (Jarboe 2001).

Castells (1993; 2000b) has identified four features that distinguish informationalism from the prior industrial stage: the driving role of science and technology for economic growth; a shift from material production to information processing; the emergence and expansion of new forms of networked industrial organization; and the rise of socioeconomic globalization.

Science and Technology

Productivity and economic growth are "increasingly dependent upon the application of science and technology, as well as upon the quality of information and management, in the process of production, consumption, distribution, and trade" (Castells 1993, 15). This is in contrast to the pre-information era, when advanced economies increased their productivity principally through infusions of capital and labor to the productive process. The importance of science and technology in economics is illustrated by the economic, and eventually political, downfall of the Soviet Union. Soviet productivity advanced regularly until 1971, simply by pumping more capital and labor into a primitive industrial system.

However, once the Soviet economy became more complex because of industrialization, it needed to rely on more sophisticated scientific processes to sustain growth. Though the Soviet Union had a great number of top-notch scientists and engineers, the overcentralized nature of the command economy made it increasingly difficult to apply science and technology to industrial processes, and the growth rates plummeted (Castells 2000a; Castells and Kiselyova 1995). In contrast, countries that were able to more flexibly integrate science and technology into the production process, such as Singapore and Korea, thrived.

Information Processing

In advanced capitalist countries there has been a shift from material production to information-processing activities, both in terms of proportion of Gross National Product (GNP) and proportion of the population employed. This entails not only a shift from manufacturing to service but also a shift within the service sector from noninformation activities (e.g., cleaning floors) to information-processing activities (e.g., computer software writing) (Castells 2000b; Reich 1991). The information-intensive industries include health care, banking, software, biotechnology, and media. But even traditional industries, such as automobile and steel, are increasingly relying on information processing in order to produce competitive products.

Networked Organization

In addition to an increasing reliance on science and technology and a shift toward information processing rather than industrial processing, there has been in recent years a shift from the standardized mass production and vertically integrated large-scale organization of the Ford era to flexible customized production and horizontal networks of economic units. In order to be able to develop, interpret, and make use of new information and knowledge as quickly and flexibly as possible, new "post-Fordist" management techniques are used that emphasize a flattened hierarchy, multiskilled labor, team-based work, and just-in-time production and distribution (Castells 1993; Gee, Hull, and Lankshear 1996; Reich 1991). Whereas the typical firm of the early twentieth century was the auto plant, with rows of assembly line workers doing a single task under orders from above, the paradigmatic firm of the early twenty-first century is the software engineering company, with teams of multiskilled employees grouping and regrouping to take on complex tasks. Indeed, this type of reorganization has even changed the automobile industry, with Toyota plants now comprising teams of multifunctional specialists rather than individual assembly line workers (Corriat, cited in Castells 2000b). Changes in work relations and production processes do not imply by any means that workplace oppression or inequalities have ended; indeed, the new economy has weakened trade unions, increased the amount of part-time work, and placed many employees on almost twenty-four-hour demand. However, employer-employee relations and employee-employee relations have taken on new forms.

Globalization

The new economy is a global one in which capital, production, management, labor, markets, technology, and information are organized across national boundaries. Global foreign direct investment grew about eightfold from 1965 to 1995, and global export of goods and services nearly quadrupled in the same period (Castells 2000b). Globalization has relied, in part, on multinational firms but also increasingly on transnational networks of firms (including both multinationals and local firms).

An example of how these four radical changes have come together to transform the economy is seen in the automobile industry. In 1977 it took about 35 person-hours of labor to assemble an automobile in the United States. New Japanese production techniques, based on technological developments and multiskilled teamwork, had brought that down to 19.1 hours by 1988 (Reich 1991). Just-in-time production and distribution techniques allowed car manufacturers to save money on inventory and warehousing, and customized, flexible global production and distribution in the 1980s gave Japanese companies an advantage over slower, more cumbersome U.S. companies (though U.S. companies eventually started to catch up). In the future, new scientific developments are expected to dramatically reduce the weight and thus engine size of cars, while increased computing power will make combustion and driving more intelligent, to the point where the value of a car will be better understood by seeing it as "chip with wheels" rather than wheels with chips (Kelly 1997, 194). And the ability to competitively design, manufacture, market, and distribute such a product internationally will be, and already is, dependent on modern telecommunications, with executives, designers, managers, and sales people around the world consulting, collaborating, communicating, and sharing information via computer-mediated networks.

The transformation of the automobile industry, and of virtually every other industry in today's world, according to the imperatives of the information economy is undeniable. And equally undeniable is the critical role of computers and the Internet in allowing these changes to take place. Though dot-coms crest and fall, the rise of "click and mortar"—existing businesses that incorporate online communication into their day-to-day functioning—is here to stay. The Internet is "transforming business practices in its relation to suppliers and customers, in its management, in its production process, in its cooperation with other firms, in its financing, and in the valuation of stocks in financial markets" (Castells 2001, 64). One illustration of this is the stunning growth of business-to-business (B2B) e-commerce, which is projected to rise in the United States alone from $400 billion USD in 2000 to $3.7 trillion USD in 2003 and to grow even faster internationally (Castells 2001, citing information from the Gartner Group).

Probably the best single example of a model "click and mortar" company is Dell Computer. Dell's ascendancy from a University of Texas college student's personal startup company in 1984 to the forty-eighth largest corporation in the United States in 2001 (with revenues higher than Microsoft, Disney, or Cisco) has sparked a cottage industry of academic and business analysis as other firms try to "Dellize."[4] The role of ICT in Dell's success, and the lessons of this for the broad information economy, are illustrated well in an excellent analysis by Kraemer, Dedrick, and Yamashiro (2000). The following discussion draws from their case study.

Dell is perhaps best known by the public for marketing its computers to consumers over the Internet. This, however, is only part of a much broader ICT-based strategic approach, which began to take shape years before the Internet surfaced as a mass phenomenon. Simply put, Dell is an information company through and through. It deploys advanced ICT in every aspect of its operations in order to gather, refine, and make instant use of customized information about its customer base, the broader market, the production process, the supply chain, distribution challenges, and service requirements. Its inventory is tiny and its manufacturing is miniscule (virtually all components are outsourced and purchased from others), but its expertise in amassing information and honing it into knowledge is the key to its profitability and success.

Information and communication technologies are key at every level of the Dell business model. Direct sales to customers take place for the most part either over the Internet or by telephone, in which case the customer service phone agent is linked directly to up-to-date inventory information and order information. Fully 70% of the company's sales are to large businesses (i.e., Fortune 1000 companies earning more than $1 million annually), and these clients order products through company Web sites that specify the menu of configurations preapproved by the particular company. These corporate customers can also get online information from Dell about their purchase histories for Dell products in specific locations, thereby enabling them to better manage and plan replacement of their computing inventory.

Sales information is immediately plugged into Dell's build-to-order system, which automatically tracks and orders inventory from suppliers to meet production demand. Just-in-time production allows Dell to maintain a cash conversion cycle of minus 8 days, which means that Dell receives money from its customers more than a week before it spends it, while assembling computers exactly to customers' specifications. Technologization of the supply, production, and distribution system has allowed the company to reduce its inventory stock from 32 days' worth in 1994 to just 6 days' worth by 1998, while cutting administrative overhead from 15% to under 10%. The production process is global, with Dell purchasing $1.6 billion in components from Taiwanese companies alone in 1998. Marketing and sales are also global, with autonomous divisions established in the United States, Europe, Asia, and Japan. The ICT units of the company are designed to be small and flexible, and are broken into smaller units whenever they reach 100 people. Finally, the company's direct contact with its consumer base allows it to market its service as well as its sales, while also providing information on rapidly changing market demand that is used in the development of new product lines.

Dell is not without its problems. Like many technology companies, its stock fell sharply in 1999–2000. Other computer companies are working hard to adopt and adapt its business model, and Dell's competitive advantage as the leading build-to-order/direct sales firm may not last. The overall market for personal computers in the United States is slowing, and it is not yet certain that Dell can reposition itself in higher-end sales and services (see "Innovator's Dilemma" in chapter 3). In sum, Dell could yet fail—and that possibility also illustrates an essential aspect of the information economy: the rapid change of fortunes of individual companies. At the same time, though, the underlying economic trends across companies are much more stable: infusion of science and technology into the production process, added value through information processing, networked forms of association and organization, and globalized marketing and production, all fueled by rapid innovations in computing and telecommunications.

Economic Stratification

Another important characteristic of the informational economy, of particular importance to issues of social inclusion, is its association with global economic stratification, both within and across countries. Both the World Bank and the United Nations Development Programme have found a sharp rise in global inequality among countries over the past forty years. The World Bank, for example, has analyzed the gap between the richest twenty countries and the poorest twenty countries over the past forty years (*World Development Report 2000/01*). In 1960 per capita Gross Domestic Product (GDP) was eighteen times that in the poorest twenty countries. By 1995, however, this gap had widened to thirty-seven times as the richest countries became much richer while the poorest countries stayed poor or became even poorer.

The United Nations Development Programme (1999b) compared the GDP of the 20% of the world's people who live in the richest countries

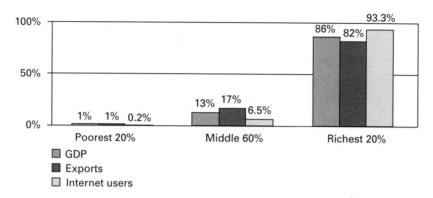

Figure 1.1
Shares of global GDP, exports, and Internet users among world's population, 1997.
Source: United Nations Development Programme (UNDP 1999b).

with the 20% of the world's people who live in the poorest countries. They found that the ratio between the two groups' GDP increased from 30 to 1 in 1960, to 60 to 1 in 1990, to 74 to 1 in 1997. By 1997, the fifth of the world's people living in the highest-income countries controlled 86% of world GDP, and the bottom fifth just 1% (figure 1.1). This corresponds to the shares of exports of goods and services received by the richest and poorest fifths and to the even sharper disparity in shares of Internet users.

This disparity of wealth, exports, and Internet use does not come about because poor countries are completely cut off from the world economy. Paradoxically, the countries of sub-Saharan Africa have a higher export/GDP ratio than developed economies do: 29 percent of GDP in the 1990s (Castells 2000b). However, exports from sub-Saharan African countries tend to be predominantly low-value primary commodities whose market value has steadily fallen in the past two decades, whereas the exports of the wealthy countries are based on high-technology and high-knowledge goods and services whose corresponding market value has steadily risen since the onset of informationalism. Between 1976 and 1996 the share of world trade composed of high- and medium-technology goods—defined as those requiring intensive research and development expenditures—rose from 33% to 54%, and the share of world trade composed of primary products fell from 45% to 24%

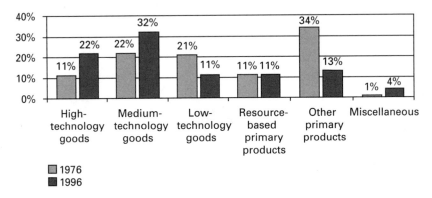

Figure 1.2
Percentage of goods in international trade by level of technology, 1976 and 1996.
Source: World Development Report 1998/99 (World Bank).

(figure 1.2) (World Development Report 1998/99). This pattern of rich countries getting richer while poor countries stay poor has resulted in the so-called twin peaks income distribution, with 2.4 billion people living in countries with average incomes of less than $1,000 USD per year, 5 billion people living in countries with average incomes of more than $11,500 USD per year, and relatively few people living in countries with average incomes of $5,000 to $11,500 USD per year (Milanovic 1999).

Global Inequality among Individuals
Global inequality among individuals is much more difficult to determine than inequalities between countries, but Francis Bourguignon and Christian Morrison (1999) have done an estimation using a measure of income inequality called the Theil index.[5] Their study shows that income inequality rose steadily through the nineteenth and early twentieth centuries, then remained at about the same level from 1920 until 1960, when it started to rise again sharply.

Branko Milanovic (1999) has carried out a study of income inequality from 1988 to 1993 based on actual household surveys of income in ninety-one countries. His data set was later reanalyzed by Yuri Dikhanov and Michael Ward (2000) and by Robert Wade (2001). Their analyses indicate that even in the short five-year period studied, global income

Table 1.2
World Income Distribution, 1988 and 1993

Inequality Measure	1988	1993	Percent Change
Richest decile's percentage of world income	48	52	8.3
Richest decile as percent of median	728	898	23.4
Poorest decile's percentage of world income	0.88	0.64	−27.3
Median as percent of poorest decile	327	359	9.8
Gini coefficient, world	63.1	66.9	6.0

Sources: Adapted from Wade (2001); Dikhanov and Ward (2000).

inequality (measured by the Gini coefficient, based on the sum total of difference from a mean) increased by 6% and the poorest decile's percentage of world income fell by 27.3% (table 1.2).

The main reason for this change seems to be a sharp distancing of the wealthiest decile from the median income; this corresponds roughly to the middle and upper classes in the wealthy countries and the elite in the poor countries—in other words, the fraction of the world's population that has been best able to profit from the information and communications technology revolution.

Inequality within Countries
Milanovic (1999) calculates that about three-quarters of the recent increase in global inequality is due to an exacerbation of intercountry differences and the one-quarter to an exacerbation of intracountry differences. Worsening inequality within countries is occurring at both ends of the spectrum, within rich and poor countries alike.

Inequality within Rich Countries   Castells (2000a) analyzed the change in income inequality in thirteen OECD countries: ten of the thirteen countries experienced a growth of internal income inequality after 1979 (table 1.3). This included not only the United States, the United Kingdom, and Australia but also traditionally egalitarian countries such as Sweden, Denmark, and Japan—all countries with a relatively high diffusion of new technologies. Castells interprets this finding as evidence

Table 1.3
Change in Income Inequality After 1979 in OECD Countries

Country	Period	Average Annual Change in Income Inequality (%)
United Kingdom	1979–1985	1.80
Sweden	1979–1994	1.68
Denmark	1981–1990	1.20
Australia	1981–1989	1.16
Netherlands	1979–1994	1.07
Japan	1979–1993	0.84
United States	1979–1995	0.35
Germany	1979–1995	0.50
France	1979–1989	0.40
Norway	1979–1992	0.22
Canada	1979–1985	−0.02
Finland	1979–1994	−0.10
Italy	1980–1991	−0.64

Source: Reprinted from Castells (2000a).

of a "structural trend toward increasing inequality in the network society" (80). This trend is grounded in the current restructuring of postindustrial economies that in turn has meant large numbers of well-paying blue-collar jobs have disappeared. In the new economy, as explained by Reich (1991), the principal division is no longer between blue- and white-collar workers but rather among three new categories: routine production workers (e.g., data processors, payroll clerks, and factory workers); in-person service workers (e.g., janitors, hospital attendants, taxi drivers); and symbolic analysts (e.g., software engineers, management consultants, strategic planners). Employees in all three categories may use computers or the Internet in their jobs, but the first two do so in routine ways (e.g., inventory checks, ordering products), whereas the last make use of ICT for analysis and interpretation of data; creation of new knowledge; international communication and collaboration; and development of complex multimedia products.

Inequality within Poor Countries   Within the handful of most impoverished countries of the world, inequality has remained relatively stable.

These countries remain outside the global ICT revolution, and the people are almost all equally poor. In some poor countries, however, economic restructuring has caused a similar polarization to that which has occurred within OECD countries. Probably the two most dramatic examples of inequality are in India and China. India has one of the largest and most developed information technology industries in the world. This industry has created a tiny group of multimillionaires and a small middle class of network and software engineers, computer programmers, and computer-assisted design specialists. At the same time, though, the benefits of the information technology revolution have had very little trickle-down effect on the country's overall population, most of which lives in desperate conditions in rural areas. The average GNP per capita in India is only $450 USD per year, 45% of adults are illiterate, and about one out of twelve children die before the age of five (*World Development Report 2000/01*).

China is a pronounced example not so much of the extent of inequality but rather of the rapid change in intracountry inequality. It has become much wealthier in recent years as the urbanized areas of the eastern seaboard (e.g., Guangzhou, Shanghai) have become integrated into the global economy. These are the same areas that have experienced rapid growth of Internet access and use, with ICT being used widely to facilitate business communication and scientific research (Foster and Goodman 2000). Nevertheless, the rural population of China remains poor, especially in the western part of the country. Gross domestic product per capita also varies in China from a high of $10,901 USD per year in Shanghai to $1,121 USD per year in Guizhou province, a ratio of 9.7 to 1 (UNDP 1999a).[6] In contrast, the ratio between per capita income in the wealthiest state in the United States to the poorest state is only 2 to 1 (United States Census Bureau 1995). This market income disparity within China has deadly consequences for people in rural areas of the country, as evidenced by China's diverging under-five mortality rates: 2.1% in the cities compared with 7.1% in the countryside (UNDP 1999a).

A similar situation holds in many other industrializing countries, such as Brazil, Mexico, and Egypt. A small percentage of the population is becoming increasingly prosperous, but a majority of the people suffer inadequate access to housing, health care, education, and stable employment. The well-to-do overlap substantially with those who have access to information and communication technologies, whereas the poor almost always lack access even to telephones.

Computer-Mediated Communication and the Network Society

While the economic sphere is a critical component of social inclusion, it is not the only component. What about the impact of ICT on other aspects of life?

Simply put, the broad trends seen in the informational economy are reverberating in all aspects of society. It for this reason that we can talk about not just the information economy but also the information society, or as Castells (2000b) puts it, the network society. Networks, based on interconnected nodes, have existed in human society since its inception, but they have taken on a new life in our time as they have become information networks powered by the Internet (Castells 2001, 2). Networks have great advantages over hierarchies because of their flexibility, speed, adaptability, and resilience. In the realm of biology, from the brain to the ecosystem, networks have proven their advantages over hierarchies time and again. In the realm of human activity, though, the communicative means for large-scale networking has been absent, and networks have until recently been unable to demonstrate their advantages over hierarchical forms of organization, which until now have been the only viable means for organizing activity on a large scale (Castells 2001). This is changing with the development of computer-mediated communication and the Internet.

Computer-mediated communication was initiated in science laboratories in the 1960s, promoted by the U.S. military in the late 1960s and early 1970s, and developed further in conjunction with key U.S. research universities through the efforts of a small cadre of programmers (Hafner and Lyon 1996).[7] This development took forms such as ARPANET, BITNET, and USENET in the 1970s and 1980s and exploded throughout the world via the Internet in the 1990s.

Computer-mediated communication qualitatively changed existing forms of representing, organizing, and sharing information in four important ways.

Written Interaction

Whether in society at large (Halliday 1993) or specifically in academia (Harnad 1991) or schools (Wells and Chang-Wells 1992), language has two main functions: it allows us to interact communicatively and to construe experience, that is to "interpret experience by organizing it into meaning" (Halliday 1993, 95). Throughout human history, the interactive role of language has been played principally by speech, whereas the permanence of written texts has made them powerful vehicles for interpretation and reflection (Bruner 1972; Harnad 1991). Writing, unlike speech, can be accessed and analyzed again and again by a limitless number of people at different times. Spoken language, on the other hand, is of the moment and deeply contextualized in a way that written texts are not. As such, the real strength of writing as a reflective and interpretive medium "was purchased at the price of becoming a much less interactive medium than speech" (Harnad 1991, 42).

Computer-mediated communication bridges this difference between spoken and written language. For the first time in human history, people can interact in a rapid written fashion at a distance. That allows them to quickly exchange ideas while maintaining a record of and reflecting on their own communication. People's own interactions can thus become the basis for epistemic engagement (Warschauer 1997).

Long-Distance Many-to-Many Communication

Written interaction, while powerful in and of itself, takes on greater significance when combined with another significant change in communication brought about by the communication revolution: long-distance many-to-many communication. For thousands of years, the only forums of many-to-many communication were the village meeting, town hall, and town square. In the twentieth century, new forms of communication such as the telephone conference call and ham radio were added to the range of "interactive broadcasting" technologies available to people, but since these were based on small, bounded numbers of oral networks, they failed to have a large social impact. In contrast, many-to-many computer-mediated communication can draw thousands of people into a single discussion, and millions of people around the world are now communicating online. While this has a potentially significant impact on almost every walk of life, from business (e-commerce) to romance (online chat and dating) to politics (public debate and grassroots organizing), one of the most profound effects is in the area of scholarship. Even before the full-blown Internet explosion of the 1990s, Stevan Harnad (1991) reported how scholarly skywriting—online exchange among scholars and scientists—was starting to reshape scientific inquiry. This exchange—which can take place via personal e-mail, specialized online scholarly forums, the online posting and archiving of works in progress and prepublication offprints, and electronic journals with much faster manuscript-to-published-document turn-arounds than paper journals— is speeding up and democratizing the means of production of knowledge. A century ago, a scientific breakthrough might have gone relatively unnoticed for months or years. Today, that same discovery can be known all over the world in a short time, and other scientists can ground their own existing and future research in these new findings without having to wait for the study to be written up and published in a print journal.[8]

A Global Hypertext

Long before the development of the personal computer, Vannevar Bush predicted the problem of information overload we are currently experiencing and proposed a system for dealing with it. In a remarkably prescient article in the *Atlantic Monthly*, Bush (1945) proposed the development of an information storage and indexing device called the memex. Bush's hypothetical memex abandoned "the traditional vertical links of library catalogs and indices—in alphabetical order, each listing under or after another—and instead externalized the associative processes of the human mind, via which any given idea was in the right circumstances equidistant, in effect equi-linkable, to any and all other ideas" (Levinson 1997, 140).

Though the memex was never built, its potential is being realized many times over by the Internet. The hypertextual organization of the Internet allows a horizontal, associative connection between sources of information just as Bush proposed. But in this case, the information linked is not that in a single office or library but a rapidly expanding mega-network of hundreds of millions of source documents put up by thousands of people all over the world.

Multimedia

Until the twentieth century, drawings, photographs, and other images played a relatively minor role in printed works, with the exception of medieval manuscripts, and audio and video elements were of course absent from printed works (Bolter 1991). The twentieth and now twenty-first centuries, on the other hand, have witnessed a steady rise of the visual and audiovisual as represented by film, radio, and television. A glance at today's mainstream books and newspapers as compared with those of a half-century ago will make clear how visual elements have expanded in the realm of print as well (Kress 1998). However, it is in computer-based multimedia, such as on the World Wide Web, that the mix of textual and audiovisual elements is most advanced. Audiovisual elements on the World Wide Web represent not just an add-on to text but a changed representational mode organized increasingly on the principle of display rather than narration (Kress 1998). Audiovisual texts are potentially a very powerful representational mode because they combine the illustrative power of the visual with the interpretive and analytic power of the written word. Multimedia have already come to dominate the world of business (e.g., advertising, presentations) and are increasingly prominent in government and education (Lanham 1993). The creation of multimedia also necessitates a complex array of semiotic, technical, and design skills and understandings, and differential access to these skills and knowledge will be one important divider between the "interacting" and the "interacted" in tomorrow's economy and society (Castells 2000b, p. 405).

The development and diffusion of computer-mediated communication represents a fourth revolution in human communication, cognition, and the means of production of knowledge, similar in impact to the three prior revolutions of language, writing, and print (Harnad 1991, 39). These new forms of networked communication make possible "an unprecedented combination of flexibility and task performance, of coordinated decision making and decentralized execution, of individualized expression and global, horizontal communication, which provide a superior organizational form for human action" (Castells 2001).

More than 500 million people are already connected to the Internet ("How Many Online?" 2001), and reliable forecasts point to 1–2 billion total users by the year 2010 (Castells 2001). Being part of this network is critical not only for economic inclusion but for almost all other aspects of life today, including education, political participation, community affairs, cultural production, entertainment, and personal interaction. ICT is making possible new organizational structures for social participation, from teen chat rooms, to online dating services, to political action Web sites, to Internet-based learning. None of these have completely supplanted face-to-face forms of communication and interaction, but they complement them as essential elements of social practice. As more forms of communication, social networking, community organization, and political debate and decision making gravitate to online media, those without access to the technology will be shut out of opportunities to practice their full citizenship. Sergio Amadeu da Silva, director of electronic governance for the City of São Paulo, expressed the importance of ICT access for social equality and inclusion in today's world:

In the information society, the defense of digital inclusion is fundamental not only for economic motives or employability, but also for socio-political reasons, principally to ensure the inalienable right to communication. To communicate in the post-modern society is the power to interact with networks of information. It is not sufficient to have a free mind if our words cannot circulate like words of others. The majority of the population, on being deprived of access to communication via computer, is simply being prevented from communicating in the most flexible, complete, and extensive means. This digital *apartheid* represents a break down of a basic formal liberty of universal liberal democracy. This brings into existence two types of citizens, one group that can instantly access and interact with what others say, and one group that is deprived of that speed of communication.[9]

ICT is particularly important for the social inclusion of those who are marginalized for other reasons. For example, the disabled can make especially good use of ICT to help overcome problems caused by lack of mobility, physical limitations, or societal discrimination. Using ICT, a blind person can access documents by downloading them from the Internet and converting text to speech; a quadriplegic can pursue a college degree without leaving home; and a child suffering with AIDS can communicate with other children around the world. Sadly, though, disabled people, because of poverty, lack of social support, or other reasons, frequently lack the means to get online. In the United States, for example, only 21.6% of disabled people have home access to the Internet, compared with 42.1% of the nondisabled population (NTIA 2000). This disproportionately low rate of Internet connectivity by those who in many senses most need it, and in one of the world's most technologically advanced countries, is evidence that market mechanisms alone are not sufficient for achieving equitable ICT access.

Of course, not all forms of networking foster social inclusion. The gaping economic and social inequality of the current era has also given rise to more ominous networks. The Al Qaeda terrorist group—bringing together at various times the Egyptian Al Jihad, the Armed Islamic Group of Algeria, Abu Sayaff in the Philippines, the Islamic Movement of Uzbekistan, the Al Itihaad Al Islamiya of Somalia, and a host of other organizations—is the archetypical networked organization, deploying sophisticated computer-mediated communication and a variety of other international media and financial devices and institutions to connect its various nodes and cells (Ronfeldt and Arquilla 2001; Zanini and Edwards 2001). Similarly the Sicilian Cosa Nostra, the U.S. Mafia, Colombian cartels, Russian *mafiyas*, and Japanese *yakuza* coordinate their smuggling, gun-running, racketeering, counterfeiting, and prostitution rings in classical network fashion (Castells 2000a). While terrorist and criminal leaders themselves often come from privileged backgrounds, they prey on conditions of social exclusion to spread their wares and influence. And these conditions of social exclusion—such as widespread hunger, rampant spread of AIDS and other diseases, massive child labor, and frightening levels of sexual exploitation and abuse—are an expression of the heightened inequality and poverty of the new global economy. Simply put, global informational capitalism, left to its own devices, has torn down traditional modes of interaction and survival, leaving hundreds of millions of people in what Castells (2000a, 165) calls the "black holes of informational capitalism." These are the parts of the world that are least connected to the information society, from the shantytowns of Soweto to the *favelas* of São Paulo, from rural India to rural Appalachia.

Some would suggest that ICT is a luxury for the poor, especially in the developing world. However, with the rapid growth of the Internet as a medium of both economic and social transaction, it is in effect becoming the electricity of the informational era (Castells 2001), that is, an essential medium that supports other forms of production, participation, and social development. Whether in developed or developing countries, urban areas or rural, for economic purposes or sociopolitical ones, access to ICT is a necessary and key condition for overcoming social exclusion in the information society. It is certainly not the only condition that matters; good schools, decent government, and adequate health care are other critical factors for social inclusion. But ICT, if deployed well, can contribute toward improved education, government, and health care, too, and thus can be a multiplying factor for social inclusion.

2
Models of Access: Devices, Conduits, and Literacy

If access to information and communication technology (ICT) is critical for social inclusion in the information era, what does such access entail? The two most common models of access to new technologies are those based on *devices* and *conduits*. The insufficiency of these two models forces us to consider a third model, based on *literacy*.

Devices

The simplest, but perhaps the most limited way, to think about ICT access is as ownership of a device. In this sense, *access* is defined in terms of physical access to a computer or any other ICT device. Ownership of a computing device is clearly part of ICT access; however, device ownership does not in itself constitute complete access because full ICT access in current times also requires connection to the Internet as well as the skills and understanding to use the computer and the Internet in socially valued ways.

The device model is appealing to some people because diffusion of devices is comparatively easy and quick, compared to diffusion of conduits, content, and practices. Devices require a one-time purchase rather than a monthly fee—let alone the development of a skill—and the purchase price is often reduced through the availability of a variety of first- and second-hand models. In the United States, for example, television and radio both reached 95% saturation points within twenty years of their introduction, and their penetration rate in low-income communities and high-income communities is about the same (currently at 97+%).

The device model has several major flaws, however. First, though the price of computers is falling, the actual purchase price is only a small part of what can be considered the total cost of ownership. This includes the price of software, maintenance, peripherals, and in institutional settings, training, planning, and administration (see comments by Kling in Patterson and Wilson 2000) as well as the price of replacement hardware and software that is necessary because of corporate-planned product obsolescence. In addition, beyond the affordability of computers (or the broader computing package), other barriers will continue to play a major role in fostering digital inequality. These barriers include differential access to broadband telecommunications; differences in knowledge and skill in using computers or in attitudes toward using them; inadequate online content for the needs of low-income citizens, especially in diverse languages; and governmental controls or limitations on unrestricted use of the Internet in many parts of the world (DiMaggio and Hargittai 2001).

In other words, the presence or absence of the computing device is only a small part of the broader context that shapes how people can actually use ICT in their lives. Although a personal computer will soon be affordable to most families in developed countries, this will not in itself overcome inequality in access to ICT for social inclusion. In developing countries, the price of a computer is of course still a major obstacle toward getting online, but the other factors just mentioned are also of great importance. For example, it accomplishes little to have a computer if you don't know how to use it.

In summary, in both wealthy and poor countries, the singular concentration on the computing device itself, to the exclusion of other factors, is a shortcoming of many well-intentioned social programs involving technology. This is particularly the case, for example, in the field of education, where too often insufficient planning regarding teacher training or curriculum reform has undermined the value of investments in costly computer equipment (see chapter 5). What is at stake is not access to ICT in the narrow sense of having a computer on the premises but rather access in a much wider sense of being able to use ICT for personally or socially meaningful ends.

Conduits

Whereas a device can be acquired through a one-time purchase, access to a conduit necessitates connection to a supply line that provides something on a regular basis. In one sense, television and radio are also conduit services in that the devices are worthless without the accompanying airwaves. However, since much television and radio programming is provided free over public airways, the device model for those technologies still holds.

Examples of conduits that require ongoing payments are electricity, telephone service, and cable television. Diffusion of conduits is slower than that of devices, either because a delivery infrastructure must be established first (such as the installation of telephone lines or fiber optic cables) or because the cost of a regular monthly fee is a disincentive to access. An illustration of the slower diffusion of conduits vis-à-vis devices is seen in figure 2.1, which shows the relatively slow diffusion

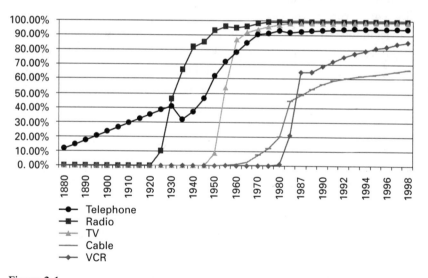

Figure 2.1
Household penetration of selected media, 1880–1998.
Source: Schement and Forbes (2000). Used with permission from *The Information Society*.

of telephones and cable television compared with television sets, radios, and video cassette recorders.

Among conduits, electricity is a useful example to consider in more detail because electricity, like ICT, has played a key role in at least one industrial revolution. Electrification has followed a variety of paths around the world based in large part on the constellation of class forces engaged in struggle over access to electric power in particular countries. In South Africa, for example, wealthy industrialists developed their electric system primarily to improve diamond, coal, and gold mining, but they did not electrify the nearby homes of their black workers (Renfrew 1984). In the Soviet Union, Lenin launched a massive national electrification effort soon after the Russian revolution under the slogan of "Communism = Soviet Power + Electrification of the Whole Country" (Abamedia 1999; Nye 1990). This campaign was largely successful, and the diffusion of electricity and power plants throughout the country was a prerequisite to the Soviet Union's rapid industrialization and eventual military success against Nazi Germany. At the same time, the highly centralized and forceful nature of the electrification campaign, as with other aspects of Soviet industrialization, took a heavy toll on the work force and citizenry.

Between these two extremes lie the experiences of Western Europe and the United States, which used different combinations of market forces and governmental action to extend universal access to electricity in the early 1900s. In the industrialized nations of Europe, strong workers' and farmers' parties pushed for electrification to be considered a social service rather than a private commodity. At the time, the state usually owned public utilities and subsequently developed electrification policies within the context of the welfare state. As a result, in Germany, Holland, and Scandinavia, 90% of all private homes and two-thirds of all farmers' homes had access to electricity by 1930 (Nye 1990). Moreover, services that grew up around the advent of electricity, such as electric trolleys, were operated by many European governments at a loss in order to provide people with affordable local transport. In the United States, however, with its weaker labor and farmer movements and its laissez-faire style of capitalism, electric utilities primarily were owned privately, with the government's role in electrification of regions reduced to regulation (Brown 1980). By 1932 public power produced only 5% of U.S.

electricity (Nye 1990). Unprofitable electric trolley systems in the United States collapsed and were replaced by privately owned automobiles. As a result of private ownership and the profit motive, American home electrification began "as a form of conspicuous consumption for the very rich, and only spread beyond the wealthiest classes at a slow pace" (Nye 1990, 140). By the end of the 1920s, 90% of U.S. farmers could not get electricity in their homes as service was not extended to their areas, and in those rural areas where service was available, farmers often had to pay twice the urban rate (Nye 1990). In the end, governmental intervention via Franklin Delano Roosevelt's 1935 Rural Electrification Act (REA) was necessary to complete the task of electrifying America. A component of Roosevelt's New Deal, the REA had its roots in decades of popular struggle for rural electrification (Brown 1980).

An interesting footnote to this background relates to the 2001 energy crisis in California. Although electric utilities are privately owned in most U.S. states, the city of Los Angeles has owned its own utilities since 1914, when a labor-backed referendum wrested the city's electric power system from private utilities. This served the city well in the energy crisis, with Los Angeles avoiding the blackouts and soaring rates experienced in the rest of California.

The lesson from these examples in terms of better understanding issues of access is not that the ICT industry needs to be government-owned; indeed, as is argued in chapter 3, privatization, done well, can be an important component of extending telecommunications access. The lesson is rather that the diffusion of any technology is a site of struggle, with access policy reflecting broader issues of political, social, and economic power.

Comparing ICT diffusion with electrification is of interest because electricity, like ICT, opened the door to a new stage of industrial capitalism. However, beyond that, the comparison no longer holds. At present, access to electricity is generally provided through a one-time infrastructure installation, with relatively small continual payments required by users and with differences of knowledge, skills, and content usually irrelevant to whether people can make use of electricity.

A closer illustrative comparison can perhaps be made between telephone service and ICT. Telephone service, like ICT, makes available an important means of public communication. Access to telephone services involves issues concerning infrastructure (e.g., telephone lines, satellites, cellular antennae) and affordability of ongoing service costs. Governments have sought to promote mass access to telephony for a number of reasons. First, and again similar to government policy concerning ICT development and access, telephone access or lack of it has been viewed as something that can help overcome or compound other disadvantages related to poverty, unemployment, and access to goods and services (Graham, Cornford, and Simon 1996). In addition, in developed countries, where telephone access has reached over 90% of the population, lack of telephone access is regarded as a direct restriction on people's opportunity to participate in societally recognized civil and social discourse (Preston and Flynn 2000). Finally, it has been recognized that universal telephone service, like universal electrification, cannot be provided by market forces alone because it involves laying expensive lines to rural areas that might have a small number of users (at least before the advent of wireless telephony) and therefore should not be subject to strict supply and demand forces of private enterprise.

Beyond the individual benefits of telephone services, governments have also promoted telephone access for reasons of collective welfare. In today's world, telephony is a key component of development, and developing countries recognize that poor and limited telephone service restricts opportunities for foreign investment and economic modernization. In addition, even in countries where many people already are connected directly to telephone services, there is a network effect to telephony; a telephone network, like a fax network, or the Internet, gains value when more users are connected to it. (Think of the alternative; only one telephone or fax machine in all the world would be valueless because no other machine would be available to receive messages). Therefore, each added network connection is seen as benefiting not only that additional user but also the entire network and society.

Despite best intentions, no government has been able to completely achieve universal telephone service, although some countries such as Finland have come quite close. In the United States, 120 years after the telephone was first introduced, only 94.2% of households have direct telephone access. The percentage with telephone access is much lower among the low-income households (78.3% for households with annual incomes below $5,000) and in households headed by African Americans (87.7%), Hispanics[1] (88.9%), or the unemployed (89.0%) (FCC 1999).

Of course, anything approaching these telephone penetration rates would be a dream in most of the developing world. Teledensity around the world, measured by number of main telephone lines per 1,000 people, ranges from 661 lines in the United States, to 364 lines in the Czech Republic, to 121 lines in Brazil, to 70 lines in China, to 22 lines in India, to 1 line in Chad (UNDP 2000). In most countries, telephone penetration trails behind television penetration, indicating the relative difficulty of achieving access to conduits versus access to devices (figure 2.2).

Chapter 3 further discusses telephone access and what kinds of policies seem to hinder or promote it. The key point to be made here is the comparatively slow and difficult diffusion of conduits, such as the Internet, as compared with devices.

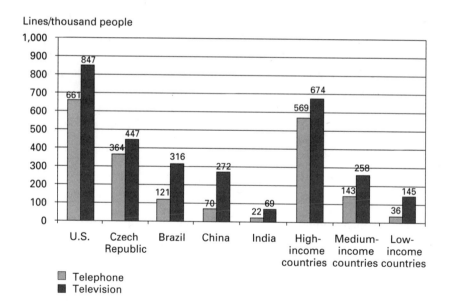

Figure 2.2
Telephone lines and television sets per 1,000 people, 2000.
Source: United Nations Development Programme (UNDP 2000).

Though conduits provide a better comparative model for ICT than devices do, neither category captures the essence of meaningful access to information and communication technologies. What is most important about ICT is not so much the availability of the computing device or the Internet line, but rather people's ability to *make use* of that device and line to engage in *meaningful social practices*. This is a crux of my argument concerning technology and social inclusion. For example, those people who cannot read, who have never learned to use a computer, and who do not know any of the major languages that dominate available software and Internet content will have difficulty even getting online, much less using the Internet productively, at least with the types of computers, Internet connections, and online content currently available.

Leah Lievrouw (2000) discusses this access issue and suggests the notion of content in contrast to conduit. Lievrouw argues that the concept of content encapsulates the physical access to a device *and* to an information channel, along with two additional elements: institutional sources of information and sufficient individual capacity to make use of that information to engage in social action and discourse. There is much of value in Lievrouw's conception of content because it moves beyond the device/conduit dichotomy. However, the usefulness of her concept is undermined by the popular, more common use of the term *content* to mean institutionally generated information. Instead, the concept of literacy more usefully provides a model because literacy, like ICT access, involves a combination of devices, content, skills, understanding, and social support in order to engage in meaningful social practices.

Literacy

There are many similarities between literacy and ICT access (table 2.1). First, both literacy and ICT access are closely connected to advances in human communication and the means of knowledge production. Second, just as ICT access is a prerequisite for full participation in the informational stage of capitalism, literacy was (and remains) a prerequisite for full participation in the earlier industrial stages of capitalism. Third, both literacy and ICT access necessitate a connection to a physical artifact (a book or a computer), to sources of information that get expressed

Table 2.1
Literacy and ICT Access

	Literacy	ICT Access
Communication stage	Writing, print	Computer-mediated communication
Main Economic Era	Industrial capitalism	Informational-capitalism
Physical artifacts	Books, magazines, newspapers, journals	Computer
Organization of content	Novels, short stories, essays, articles, reports, poems, forms	Web sites, e-mail, instant messages
Receptive skills	Reading	Reading and multimedia interpretation, searching, navigating
Productive skills	Writing	Writing and multimedia authoring and publishing
Divides	A great literacy divide?	A digital divide?

as content within or via that physical artifact, and to a skill level sufficient to process and make use of that information. Fourth, both involve not only receiving information but also producing it. Finally, they are both tied to somewhat controversial notions of societal divides: the great literacy divide and the digital divide.

To fully understand the relationship, it is worth exploring in more depth what literacy is, how it develops, and what research has shown regarding the existence of a literacy divide.

The Practice of Literacies

While the commonsense definition of *literacy* is the individual skill of being able to read and write, "new literacy" theorists prefer a broader definition that takes into account the social contexts of literacy *practice*.[2] They point out that what is considered skillful reading or writing varies widely across historical, political, and sociocultural contexts (Gee 1996). For example, in the pre-Gutenberg era, writing principally involved memorizing and transcribing oral speech or carefully and accurately copying classical or religious manuscripts (McLuhan 1962). A skilled writer in those times usually had outstanding mnemonic and penmanship abilities. Reading was often done publicly, with an orator slowly reading a manuscript out loud to a gathered group. Whether done publicly or privately, however, the purpose of reading generally was to interpret a small number of classical and religious texts in order to achieve "a new consciousness of what a text *could have meant* or *could mean* to a putative reader" (Olson 1994, 157, emphasis in original).

These notions of reading and writing started to shift as early as the twelfth century (Olson 1994) but changed much more rapidly following the introduction of the printing press in the mid-fifteenth century. In this new typographic era, scholarly writing came to be viewed as authorship of original material, and scholarly reading came to mean the gathering, comprehending, and making use of information from a variety of sources (Eisenstein 1979).

Notions of literacy have continued to change in the past 100 years. For example, Suzanne de Castell and Allan Luke (1986) identify three distinct paradigms of school-based literacy in recent U.S. history, each highly dependent on the social, economic, and cultural norms of particular epochs. First, in the nineteenth-century classical period, literacy was viewed in terms of knowledge of literature and attention to rhetorical appropriateness. Literacy pedagogy involved rote learning, oral recitation, copying, and imitation of what was considered correct speech and writing. And the literacy curriculum was based on exemplary texts such as the Bible, a narrow selection from Greek and Roman literature, and handwriting primers. This public schooling paradigm corresponded to the needs of an aristocratic social structure, in which land, power, and knowledge was concentrated in few hands, and education involved obedience to tradition and power.

Following the mass industrialization of the early twentieth century, a Deweyan progressive paradigm of literacy emerged as a "self-conscious attempt . . . to provide the skills, knowledge, and social attitudes required for urbanized commercial and industrial society" (de Castell and Luke 1986, 103). In this paradigm, literacy was viewed as a form of self-expression. Literacy pedagogy involved teacher/pupil interaction and the "discovery method." The literacy curriculum included civics in order to produce good citizens, adventure stories in order to tap into students' interests, and self-generated written texts in order to foster creativity and imaginative thinking.

But the progressive model never fully took hold; instead, it was in constant struggle with a more technocratic paradigm that eventually won out (Cuban 1993). Within this technocratic paradigm, literacy was viewed in terms of the skills needed for functioning effectively in society. Literacy pedagogy involved programmed instruction, learning packages with teacher as facilitator, and mastery learning of a common set of objectives. And the literacy curriculum was based on decontextualized subskills of literacy competence.

From this brief historical sketch, we can conclude that literacy is not a context-free value-neutral skill; rather, being literate "has always referred to having mastery over the processes by means of which culturally significant information is coded" (de Castell and Luke 1986, 374). For this reason, the plural form *literacies* is often used by literacy theorists. In the same vein, scholars often prefer to use the term *literacy practices* rather than *literacy skills* because the former term emphasizes the application of literacy in a social context rather than as a decontextualized cognitive ability.

The Literacy Divide

One of the most important theoretical questions related to literacy, and one that corresponds closely to current debates over a digital divide, is whether there exists a great literacy divide. Literacy is distributed and practiced on a highly unequal basis. Adult literacy rates range from over 99% in some of the most developed countries (including Italy, Spain, Israel, Singapore, Greece, and South Korea), to the 50%–60% range in some developing countries (e.g., 55.7% in India, 53.7% in Egypt), to under 30% in some of the poorest countries (e.g., 22.2% in Burkina Faso, 14.7% in Niger) (UNDP 2000). Literacy is highly correlated with income and wealth at both the individual and societal levels. So, in one sense, the importance of literacy in social and individual development is broadly recognized.

What is disputed is the issue of causality, that is, whether literacy enables development, or whether unequal development (and corresponding unequal distribution of political, economic, and social power) restricts people's access to literacy. Some advocates of the former notion posit the existence of a literacy divide. From this perspective, there are fundamental cognitive differences in individuals who are literate and who are not, resulting in a great literacy divide at both the individual and societal levels. Literacy has been said to separate prehistory from history (Goody and Watt 1963), primitive societies from civilized societies (Lévi-Strauss, in Charbonnier 1973), and modern societies from traditional societies (Lerner 1958; see discussion in Scribner and Cole 1981). At the individual level, literacy has been said to allow people to master the logical functions of language (Goody 1968; Olson 1977) and to think abstractly (Greenfield 1972; Luria 1976).

The imputed cognitive benefits of literacy have proven difficult for researchers to investigate. The problem is that literacy is almost always confounded with other variables, particularly with schooling. For the most part, those who are completely illiterate tend to have had little or no schooling, whereas those with high levels of literacy tend to have had a good deal of schooling.[3] And amount of schooling usually correlates directly with income levels of a child's family or the work engaged in by the child's family. The covariance of literacy with other social factors such as schooling and family employment has made the cognitive impact of literacy a thorny focus of investigation.

Two educational psychologists, Sylvia Scribner and Michael Cole, developed a creative solution to this research problem. They identified a tribe in Liberia, the Vai, that had developed its own written script in the tribe's own local language. Literacy in the Vai script was passed on through informal tutoring, not through formal schooling. Vai writing was used in very limited ways, mostly for personal correspondence and business records. By carrying out a three-way study that compared illiterate tribal members, those literate only in the Vai language (through personal tutoring), and those with broader English or Arabic literacy skills gained through schooling, Scribner and Cole (1981) were able to separate which cognitive benefits could be most likely attributed to literacy and which others were most likely due to the broader environment of formal education.

Interestingly, Scribner and Cole found virtually no generalizable cognitive benefits from Vai literacy. Individual differences on a range of cognitive tasks, in areas such as abstraction classification, memory, and logic, were instead due to other factors, such as schooling or, in some cases, living in an urban (rather than a rural) area.

Vai literacy was found to be correlated with better meta-language understanding when compared to non-Vai literate people's meta-language understanding. For example, Vai literates were better able than nonliterates to provide grammatical explanations of an oral sentence, read pictures (decode graphics according to a preassigned code), and write with pictures. Similarly, the cognitive benefits of Arabic literacy seemed to be closely associated with the functions of its use. The main benefit of Arabic literacy was in the area of verbal recall, which is not surprising because Arabic literacy is developed in Liberia through memorization of the Koran. More complex and generalizable cognitive tasks, such as solving abstract logic problems, were correlated only with schooling and English literacy, which is, again, not surprising, given the types of abstraction and problem solving that are practiced in school. And on no single task in their entire study did every Vai literate outperform every nonliterate (in other words, individual variation trumped group variation according to literacy level).

Scribner and Cole's study helped settle the question of whether there is a great literacy divide, at least at the individual level. Their work showed that there is no single construct of literacy that divides people into two cognitive camps. Rather, there are gradations and types of literacies, with a range of benefits closely related to the specific functions of literacy practices. Literacy, in a general sense, cannot be said to cause cognitive or social development; rather literacy and social development are intertwined and co-constituted, as are technologies and society in general (see chapter 7).

Acquisition of Literacy

If literacy is understood as a set of social practices rather than as a narrow cognitive skill, this has several important consequences for thinking about the acquisition of literacy, and important parallels with the acquisition of access to ICT. Literacy acquisition, like access to ICT, requires a variety of resources. These include physical artifacts (books, magazines, newspapers, journals, computers); relevant content transmitted via those artifacts; appropriate user skills, knowledge, and attitude; and the right kinds of community and social support. Let us examine these one at a time, in reference to literacy.

First, the physical artifacts available for reading and writing enable the acquisition and practice of literacy at both the individual and the societal levels. At an individual level, it has long been known that children have an easier path to reading in a text-rich environment, and the ready availability of books and other reading materials is a valued part of effective reading programs (Krashen 1989). At the societal level, the mass production of books—and eventually newspapers, magazines, and periodicals—following the development of the printing press was critical for the achievement of mass literacy (together with other factors; see chapter 7).

Equally important for acquisition of literacy is relevant content within or via these books and artifacts, in terms of language, level, topic, and genre. One of the major obstacles to literacy acquisition is the dearth of published material in many if not most of the 7,000 languages that are spoken around the world. In addition, Paolo Freire (1994) and others have shown that literacy instruction is most effective when it involves content that speaks to the needs and social conditions of the learners. And, as with ICT-related material (see chapter 4), this content is often best developed by the learners themselves. Content that is relevant to people's lives is critical not just for basic literacy but for all levels; graduate students could never learn to interpret scientific articles, much less to write them, without direct access to relevant scientific writing.

Third, literacy acquisition requires the development of a variety of skills, knowledge, and attitude. Cognitive processing skills are required at both the bottom-up level (e.g., word recognition) and top-down level (e.g., guessing words and their meaning from context). And while we often focus on the skill of reading, knowledge and attitude are equally important. Reading is a transitive verb; learning to read inevitably means learning to read *something* (see discussion in Gee 1996). And to read and understand that something involves bringing to bear a vast amount of background knowledge, or schemata. Understanding a simple article about a basketball game involves a wealth of background knowledge about the way the game is played, who the teams and players are, how sports are typically reported in newspapers, and even how a newspaper is laid out. Attitudes involve the motivation and desire to read, the level of confidence in reading, and the general disposition to read various kinds of texts in different ways.

Finally, and most important, learning to read is a social act that intersects with social structure, social organization, and social practices. As Gee (1996, 41) explains, "A way of reading a certain type of text is *only* acquired . . . by one's being embedded (apprenticed) as a member of a social practice wherein people not only read texts of this type in this way, but also talk about such texts in certain ways, hold certain attitudes and values about them, and socially interact over them in certain ways."

This can be illustrated by referring to two very different types of literacy practices: those of a Pakistani *madrassa* (religious school) and those of an American university. The literacies valued in a *madrassa* involve, among other things, memorizing texts of a certain length to be repeated or referred to in certain ways in very particular contexts, and reciting appropriate passages in proper ways in the appropriate context. Is there any doubt that these can only be acquired through interaction, discussion, learning, and religious practice in social context with other Muslims in very particular settings? Similarly, consider the amount and types of social engagement that enable American students to learn to read and write in ways that their professors value, a process described nicely by Bartholomae (1986, 4): "Every time a student sits down to write for us, he has to invent the university for the occasion. . . . He has to learn to speak our language, to speak as we do, to try on the peculiar ways of knowing, selecting, evaluating, reporting, concluding, and arguing that define the discourse of our community."

Finally, the multifaceted nature of literacy, the range of resources it requires, and the social nature of its practice and mastery all point to one inevitable conclusion: the acquisition of literacy is a matter not only of cognition, or even of culture, but also of power and politics (Freire 1970, 1994; Freire and Macedo 1987; Gee 1996; Street 1984, 1993, 1995). From South Africa to Brazil to the impoverished ghettos of the

United States, access to literacy intersects with unequal opportunities to attend school, inequitable distribution of resources within the educational system, and curricula and pedagogy that meet the needs of certain social groups more than others. Perhaps the most obvious evidence of this phenomenon is the appallingly low rate of women's literacy in many countries in the world today, such as Burkina Faso (13.3% for females; 33% for males), Nepal (22.8% for females; 58% for males), or Morocco (35.1% for females; 61% for males).[4] Because of the politicized nature of literacy, campaigns that focus exclusively on individual skill while ignoring broader social systems that support or restrict extended literacy are not always the most effective. In many cases literacy is not so much granted from above as seized from below through the social mobilization and collective action of the poor and dispossessed.

Literacy and ICT Access

A synthesis of the previous discussion yields six principal conclusions about literacy:

• There is not just one type of literacy, but many types.

• The meaning and value of literacy varies in particular social contexts.

• Literacy capabilities exist in gradations rather than in a bipolar opposition of literate versus illiterate.

• Literacy alone brings no automatic benefit outside of its particular functions.

• Literacy is a social practice, involving access to physical artifacts, content, skills, and social support.

• Acquisition of literacy is a matter not only of education but also of power.

These points serve well as the basis for a model of ICT access: There is not just one type of ICT access, but many types. The meaning and value of access varies in particular social contexts. Access exists in gradations rather than in a bipolar opposition. Computer and Internet use brings no automatic benefit outside of its particular functions. ICT use is a social practice, involving access to physical artifacts, content, skills, and social support. And acquisition of ICT access is a matter not only of education but also of power.

The development of a more sophisticated understanding of literacy did not lead to downplaying its importance. Rather, by better understanding literacy, academics, educators, and policymakers could better promote it. Similarly, by better understanding the broad and complex nature of ICT access, we can also better promote it. Access to ICT for the promotion of social inclusion cannot rest on providing devices or conduits alone. Rather, it must engage a range of resources, all developed and promoted with an eye toward enhancing the social, economic, and political power of the targeted clients and communities. Any attempt to categorize these resources is by nature arbitrary, but an analysis based on four general categories serves the purposes of both analysis and policymaking. These categories have emerged from my ethnographic research in Hawai'i (Warschauer 1999) and Egypt (Warschauer 2001a) as well from my case study research in California, Brazil, and India. They have been identified in similar terms by other researchers and theorists who have examined issues of technology and social inclusion in various contexts (e.g., Aichholzer and Schmutzer 2001; Carvin 2000; Wilson 2000). They can be labeled physical resources, digital resources, human resources, and social resources (figure 2.3). Physical resources encompass

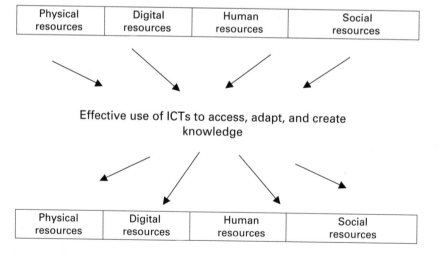

Figure 2.3
Resources contributing to ICT access.

access to computers and telecommunication connections. Digital resources refer to digital material that is made available online. Human resources concern issues such as literacy and education (including the particular types of literacy practices that are required for computer use and online communication). Social resources refer to the community, institutional, and societal structures that support access to ICT.

In considering these four sets of resources, it is important to realize their iterative relation with ICT use. On the one hand, each resource is a *contributor* to effective use of ICTs. In other words, the presence of these resources helps ensure that ICT can be well used and exploited. On the other hand, each resource is a *result* of effective use of ICTs. In other words, by using ICTs well, we can help extend and promote these resources. If handled well, these resources can thus serve as a virtual circle that promotes social development and inclusion. If handled poorly, these elements can serve as a vicious cycle of underdevelopment and exclusion.

3

Physical Resources: Computers and Connectivity

Although full access to information and communication technology (ICT) requires more than just the presence of devices and conduits, there still remain pressing issues concerning physical access to computers and the Internet. Examining data that indicate who does and does not have physical access reveals a number of interesting trends, as does an analysis of strategies and approaches put in place to enhance people's physical access through more affordable computers, Internet access, and public access centers.

Who Is Connected?

As of August 2001 an estimated 513 million people around the world had Internet access ("How Many Online?" 2001). That represents some 8.4% of the world's people.[1] Even though Internet access has been increasing rapidly in some developing countries, access remains highly stratified by region. The number of people with Internet access—defined as those who have been online in the last three to six months—ranges from 57.2% in North America to 0.5% in Africa (table 3.1 in "How Many Online?").

The reasons for disparity in Internet access rates are multiple and involve issues of economics, infrastructure, politics, education, and culture. Several studies have been conducted that analyze the principal factors that correlate with differential Internet access rates. One of the largest, conducted by Kristopher Robison and Edward Crenshaw (2000), examined the interrelationship between number of Internet host computers per capita and several economic, social, and political variables in

Table 3.1
Internet Access Rates, August 2001

Region	No. of People with Internet Access (millions)	Percentage of Population with Internet Access
U.S. and Canada	181	57.2
Europe	155	21.3
Latin America	25	4.8
Asia	144	3.9
Middle East	5	2.4
Africa	4	0.5
World	513	8.4

Source: Adapted from "How Many Online" (2001); Population Reference Bureau (2001).

seventy-five developed and developing countries. They found that the strongest factor correlating with Internet access is teledensity; countries with more telephone lines per 1,000 people also tend to have greater Internet access. Other factors, in order of importance, include high levels of economic development (measured by energy consumption), a post-industrial economy (measured by size of the service sector), educational level (measured by number of secondary students per population), and political openness (measured by a composite of factors that include elective government and constitutional constraints on governmental power).

Another study, conducted by Ezster Hargittai (1999) examined international variation in Internet connectivity among eighteen member countries of the Organization for Economic Cooperation and Development (OECD). These countries are all relatively developed industrially, yet Internet penetration among them ranges from 881 hosts per 10,000 inhabitants in Finland to only 26 hosts per 10,000 inhabitants in Greece (figure 3.1).

As in the Robison and Crenshaw study, telecommunications was found to be the most important factor in terms of Internet access rates. In Hargittai's study, telecommunications was measured according to teledensity rate and the existence of a competitive (rather than monopolized)

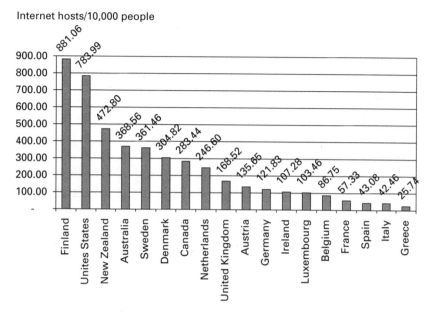

Internet hosts/10,000 people

Finland	881.06
Unites States	783.99
New Zealand	472.80
Australia	368.56
Sweden	361.46
Denmark	304.82
Canada	283.44
Netherlands	246.60
United Kingdom	168.52
Austria	135.65
Germany	121.83
Ireland	107.28
Luxembourg	103.46
Belgium	86.75
France	57.33
Spain	43.08
Italy	42.46
Greece	25.74

Figure 3.1
Internet host distribution in selected OECD countries, 1998.
Source: *Telecommunications Policy* 23 (10–11), "Weaving the Western Web: Explaining Differences in Internet Connectivity among OECD Countries," by Eszter Hargittai. Copyright 1999. Used with permission from Elsevier Science.

telecommunications industry. Other factors, in order of importance, included English language proficiency (measured by number of secondary students studying English), education (measured by combined primary, secondary, and tertiary enrollment rates), national wealth (measured by Gross Domestic Product, GDP), and equality (measured by the Gini coefficient).

Finally, the OECD has gathered information on the relationship of Internet access costs and Internet host density as a measure of accessibility. While there is not a one-to-one correspondence between the two variables, in general the OECD study found there is a strong correlation between cost and density (figure 3.2).

Without question, there are a wide variety of variables associated with Internet penetration. Given this context, the question remains as to what countries can do to promote greater physical access to computers and

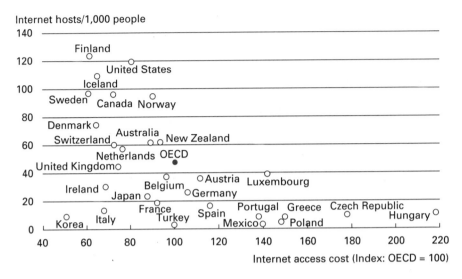

Figure 3.2
Internet access costs and Internet ost density, 1999.
Source: Organization for Economic Cooperation and Development (OECD 2000). Used with permission.

the Internet. A close examination of responses made by developed and developing countries to issues pertaining to computer and Internet access reveals much about the tenaciousness of inequality where ICT is concerned, and suggests that developed and underdeveloped countries each face very different challenges in helping promote ICT access.

Physical Access in Developed Countries

As shown in figure 3.2, there is a great deal of disparity in Internet access rates among developed countries. Hand-in-hand with this disparity comes the vexing issue of inequality *within* developed countries, with access to the Internet generally stratified by socioeconomic status and race. Consequently, developed countries face two specific and inter-related challenges regarding physical access to the Internet: the need to improve the *quantity* of access (increasing the general level of Internet penetration in the country) and the need to provide *more equal access* among their citizens.

To illustrate this point, it is useful to look at some examples from developed countries. As Hargittai (1999) has pointed out, Finland occupies a striking position as being the "most wired" nation. There are several reasons for this, including Finland's competitive telecommunications sector; its correspondingly low Internet access charges, its flat-rate Internet access charges (at least during off-peak hours, in contrast to the per minute charges in many countries); its high per capita information technology production; its high level of English (the dominant language of the Internet; see chapter 4); and the country's implementation of a ground-breaking national information society strategy in 1994 (Hargittai 1999).

In contrast, countries such as Greece, Italy, Spain, and France all have very low rates of Internet access. Particularly striking is the great disparity between Internet access in Finland and France, because the two countries have approximately equal GDP per capita (*World Development Report 2000/01*). Possible explanations for this difference include the less open and less competitive nature of France's telecommunications industry in comparison to that of Finland and higher Internet access costs in France (OECD 2000; 2001). In addition, fewer French people speak English than do people in Northern European countries like Finland ("Languages in Europe" 2002). And France has never embraced the Internet as ambitiously as Finland, partly because of its alternative and competing Minitel system. The difference between these two countries indicates that economic wealth is far from the only factor determining Internet access. Indeed, Italy has an even lower rate of Internet access than France, despite having a GDP approaching those of France and Finland. In developed countries (as well as in developing countries) telecommunications policy appears to be the major factor affecting overall physical access to the Internet, with competition, low rates, and teledensity all correlated with Internet access rates.

A second issue in developed countries is that of unequal access. This calls for an examination of the extent to which physical access to the Internet mirrors and contributes to other levels of inequity. Not surprisingly, this issue has received the most attention in the United States, the country where the Internet was first launched and where socioeconomic disparities are quite pronounced compared to other developed countries.

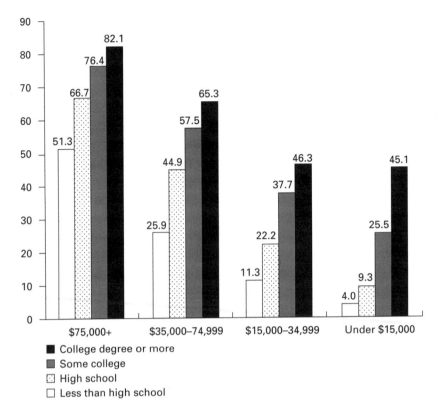

Figure 3.3
Percentage of U.S. households with Internet access, by income and education, 2000.
Source: National Telecommunications and Information Administration (NTIA 2000). Used with permission.

The most authoritative studies to date on this issue have been conducted by the U.S. Department of Commerce Telecommunications and Information Administration, in a series of *Falling Through the Net* reports. The latest of these reports (NTIA 2000) indicates several significant trends regarding computer and Internet access in the United States.

First, computer and Internet access remains highly stratified by race, income, and education. Income and education appear to be the dominant stratifiers; for example, high-income college graduates have Internet access rates over fifteen times higher than low-income high school dropouts (figure 3.3).

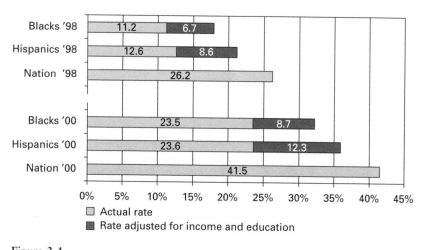

Figure 3.4
U.S. Internet access, contribution of income and education differences to racial gap, 1998 and 2000.
Source: National Telecommunications and Information Administration (NTIA 2000). Used with permission.

Second, race and ethnicity are themselves stratifying factors, exclusive of their intersection with income and education. Figure 3.4 shows that income and education account for only about half the gap between the national average and blacks' and Hispanics' access to computers and the Internet. This can be explained by a number of factors. For example, blacks and Hispanics have not only lower incomes and education but also lower assets and savings, lower literacy rates, and fewer personal connections with others who know how to use computers.

Third, Internet access among high-income groups is starting to reach saturation levels, with more than two-thirds of all households earning more than $50,000 USD connected to the Internet. On the other hand, Internet access among marginalized groups—whether by race, income, education, or geographic location—is growing the fastest (table 3.2).[2] This indicates that the classic S-shaped curve is occurring in the United States, with the early adopters and early majority now having access and the late majority starting to gain access now.[3] By a similar process, disparity by gender has already been overcome in the United States, with some 44.6% of men and 44.2% of women using the Internet by August 2000.

Table 3.2
Percentage of U.S. Households with Internet Access, 1998 and 2000

	Households (%)		Point Change	Expansion Rate (%)
	December 1998	August 2000		
Ethnic Group				
All	26.2	41.5	15.3	58.4
Black non-Hispanic	11.2	23.5	12.3	109.8[a]
Hispanic	12.6	23.6	11.0	87.3[a]
White non-Hispanic	29.8	46.1	16.3	54.7
Asian Amer. and Pacific Islander	36.0	56.8	20.8	57.8
Income				
Less than $15,000	7.1	12.7	5.6	78.9[a]
$15,000–$24,999	11.0	21.3	10.3	93.6[a]
$25,000–$34,999	19.1	34.0	14.9	78.0[a]
$35,000–$49,999	29.5	46.1	16.6	56.3
$50,000–$74,999	43.9	60.9	17.0	38.7
$75,000 and above	60.3	77.7	17.4	28.9
Education				
Less than high school	5.0	11.7	6.7	134.0[a]
High school graduate	16.3	29.9	13.6	83.4[a]
Some college	30.2	49.0	18.8	62.3[a]
College graduate	46.8	64.0	17.2	36.8
Postgraduate	53.0	69.9	16.9	31.9
Geographical Location				
Rural	22.2	38.9	16.7	75.2[a]
Urban	27.5	42.3	14.8	53.8
Central city	24.5	37.7	13.2	53.9

Source: Adapted from National Telecommunications and Information Administration (NTIA 2000).
[a] Above the average 58.4% expansion rate.

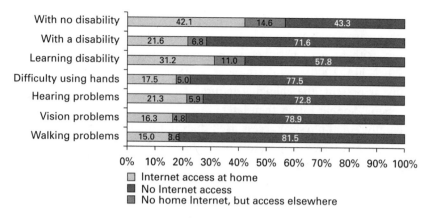

Figure 3.5
U.S. Internet access, by disability status, 1999.
Source: National Telecommunication and Information Administration (NTIA 2000). Used with permission.

Fourth, stark differences exist in access rates between the disabled and nondisabled. As mentioned earlier, only 21.6% of the disabled have home access to the Internet compared with 42.1% of people without disabilities. The discrepancy is even greater when rates of home access and access elsewhere are combined: 28.4% for disabled versus 56.7% for nondisabled (figure 3.5). All told, while just under 25% of people without a disability in the Untied States have never used a computer, close to 60% with a disability fall into that category.

In spite of the fast growth in Internet access among some disadvantaged groups, unequal physical access to the Internet in the United States is likely to remain a long-term concern. First, prior experience with other conduit services (such as telephone service) suggests that the expansion of the Internet will likely leave out at least a small percentage of the population for decades to come, and that this group will heavily overlap with those who are marginalized in other ways, such as by education, income, or disability. Second, even if discrepancies of access to basic Internet services are eventually overcome, the highly innovative nature of the computing and telecommunications industries indicates that new forms of technological disparities will arise in the near future. Research suggests that, in general, blacks and Hispanics, the poor, those with limited education, and those living in female-headed households are less likely to have advanced features at home such as broadband access, faster computers with modern peripherals, or multiple home computers (Becker 2000; NTIA 2000). While these services might seem like luxuries to many at present, the changing nature of the Internet may make them necessities sooner rather than later.

Finally, equal physical access itself does not imply equal ability to use ICT, which is affected by factors other than physical access, such as education and literacy (see chapter 4), content and language (see chapter 5), and social capital (see chapter 6).

Physical Access in Developing Countries

Physical access to the Internet is at a much lower rate in developing countries. Whereas wealthy countries face the challenge of providing universal service (ensuring everybody has the opportunity to have Internet service in the home), developing countries face the more limited but nonetheless pressing goal of providing universal access (ensuring that everybody has the opportunity to make use of the Internet somewhere—at home, work, school, a community technology center, or a rural telecenter). In terms of examining key challenges faced by developing countries in providing Internet access, three countries—Egypt, India, and China—provide useful examples.

Egypt

The Internet first entered Egypt in 1993, when a small university network was established in Cairo (*Information Technology in Egypt* 1998). The following year, the Egyptian cabinet's Information and Decision Support Center (IDSC) and the Regional Information Technology and Software Engineering Center (RITSEC) introduced the Internet more broadly, offering free Internet services for corporations, government agencies, nongovernmental organizations, and professionals. The Internet was commercialized in Egypt in 1996 and since then has grown rapidly, from about ten Internet service providers (ISPs) and some 20,000 users in 1996 to an estimated one hundred ISPs and 400,000 users in the year 2000, representing about 0.7% of the population.

The growth of the Internet in Egypt is constrained by many factors, the principal one economic. With a GDP per capita of $120 USD per month, few families can afford the $500+ USD costs involved in purchasing a computer, and fewer still can afford the flat rate of $15 USD per month Internet access charges. The economic strain is compounded by the fact that local telephone calls cost $1.00–$3.00 USD an hour, making frequent Internet use expensive even for the Egyptian middle class.

A second barrier is teledensity. At present, there are only about six phone lines per 100 people in Egypt, with less than two lines per 100 people in rural areas, which clearly places further restrictions on people's ability to access the Internet.

A third factor is literacy and education. Only 53.7% of the adult population in Egypt is literate, and few who are literate know how to use a computer.

A fourth factor is language skills. There is very little Internet content in Arabic, and most Internet use and computer-mediated communication in Egypt is in English. Only a small elite knows English well enough to use it on a regular basis for online research or communication, and few options exist yet for Arabic-language Internet use. (See additional discussion of language issues in chapter 4.)

Finally, uneven development and unequal distribution of resources also affect the growth of online access. Most of the wealth and resources of Egypt are concentrated in a few urban centers, principally Cairo and Alexandria. Most ISPs are also located in these cities. Egyptians living elsewhere usually have to pay long-distance connection rates, thus making Internet access even more expensive.

The bottom line is that Internet access is economically beyond the reach of Egypt's rural and urban poor, and can even become expensive for the country's middle class. People who log on cope with these expenses in a variety of ways, such as sharing Internet accounts or accessing the Internet via cyber cafés. Egypt is purported to have one of the highest rates of shared Internet accounts in the world, and the estimated total of 400,000 users throughout the country is based on only about 60,000 accounts. However, even beyond these individual and national economic factors, factors related to language, content, and education

Table 3.3
Internet and Mobile Phones in Egypt, 2001

	Internet	Mobile Telephones
First introduced	1994	1998
No. of users (2001)	400,000	2,000,000
Device	Computer (expensive, not very portable)	Mobile telephone (inexpensive, portable)
Monthly cost/ Basic account	~$15 USD (plus per minute charges for telephone use)	~$20 USD (plus per minute charges for telephone use beyond basic minutes)
Language	Principally English	Principally Arabic
Skill required	High skill level	No special skills

also hold down Internet use. For example, though mobile telephony can cost as much as or more than an Internet account, and mobile telephones were introduced five years later than the Internet in Egypt (in 1998), there are now some 2 million mobile phone users in the country, or nearly five times the number of Internet users. This outcome is the result of a number of causes, including, for example, the ease of use of a mobile telephone compared to the Internet and the fact that telephone communications can be conducted in Arabic instead of English (table 3.3).

India

No developing country has a more prominent information and communication technology sector than India, but actual access to the Internet reaches only a tiny fraction of the Indian population. On the one hand, the city of Bangalore alone has 300 ICT companies, including some world leaders in software and Internet design. On the other hand, 45% of the population cannot read or write, 44% of the people survive on less than $1 USD a day, and 370,000 villages in the country lack telephone connections (Hanson 2001; *World Development Report 2000/01*).

The Internet was first introduced in India in 1995. The number of users since then has risen from 10,000 that year to an estimated 5.5 million by 2001, which nonetheless represents only 0.5% of the population. As in Egypt, the use of the Internet in India is largely restricted to major urban areas; some 85% of users are found in eight large cities. In addition, 63% of users are under the age of 35, and 72% of users are male (Hanson 2001).

Factors restricting broader use of the Internet in India are similar to those in Egypt: low teledensity (2.2 per thousand) (*World Development Report 2000/01*); high rates of poverty; and limited knowledge of English outside of a small elite, combined with an as yet limited presence of Indian languages online.

China

China has the largest number of Internet users among developing countries and is expected within the next two decades to have the largest number of any country in the world. China first introduced the Internet in 1994 and now has some 26.5 million users, representing about 2% of the population (CNNIC 2001). The Internet in China has a broader geographic base than in India or Egypt, with network services reaching 230 cities throughout the country (Foster and Goodman 2000). However, there is still a strong geographic concentration of ICT service provision. The four wealthy Eastern areas of Beijing, Shanghai, Shandong, and Guangdong between them have some 50% of all the users in the country, whereas the impoverished Western provinces of Tibet, Qinghai, and Ningxia have less than 0.3% of the nation's Internet users among them (CNNIC 2000). Some 98% of Internet users in China have at least two years of college, a status achieved by only a small minority of Chinese people. As in India, the majority of regular Internet users in China are young (80% are 35 or under) and male (61%) (CNNIC 2001). Since most computing in China is done in Chinese—a language with a large number of Web sites—language does not pose the same barrier to Internet use as it does in Egypt or India.

While there is of course a great deal of diversity with regard to Internet access in developing countries, a number of significant patterns can be identified within the cases of India, China, and Egypt that appear to be typical for much of the developing world. First, Internet access only reaches a tiny percent of the population. Second, use is highly stratified and concentrated among a predominantly male, young elite based in major cities. Third, while there remains quite a bit of room for growth, none of these countries can expect to provide Internet service to the majority of citizens in the coming decades. Persistent problems of low teledensity, high poverty, rural underdevelopment, and limited education mean that the policy goal in the near future must be expanded service and universal access rather than universal service per se.

What Is to Be Done?

With these cases as a backdrop, efforts to achieve greater physical access in developed and developing countries are thrown into high relief. In particular, three issues are foregrounded in any analysis of formal moves to enhance physical access to ICT: affordability of computers, extension and affordability of telecommunications, and the provision of public access centers.

Affordability of Computers

The price of an entry-level computer has fallen steadily over the last several years and is expected to continue to fall in the coming years (table 3.4). For example, in the United States, the estimated entry-level computer price for 2002 is $577, an amount that is starting to approach what the average family might pay for a television set (Thierer 2001).

Given the high rate of innovation in the computer industry, the rapidly falling ratio of cost per computing power, the steadily decreasing cost of entry-level computers, and the continuous expansion of computers into

Table 3.4
Average Price of Base Model Personal Computer in the United States, 1996–2002

Year	Price (USD)	Annual Change (USD)
1996	$1,747	—
1997	$1,434	−$313
1998	$1,139	−$295
1999	$916	−$223
2002 (est.)	$577	

Source: Adapted from Thierer (2001).

larger numbers of households, there is not much to be gained by large-scale governmental provision of home computers in developed countries. Indeed, such programs can be counterproductive, either by draining resources from potentially more beneficial governmental initiatives, such as training programs, or by locking in particular hardware (or networking) infrastructures that end up slowing down innovation (e.g., the Minitel system in France has apparently slowed down the spread of more broad-based Internet service and access, thus undermining its laudable goal of providing socially relevant services online).

While general, large-scale governmental intervention in provision of computers is most likely unnecessary and perhaps even unhelpful, providing computers in particular circumstances tied to specific uses can of course be helpful. One example is "learn-to-earn" programs, whereby low-income participants in community development programs eventually get to keep a computer after going through a substantial amount of formally organized training (Clines 2001). Other examples include schools and school systems, where the educational goals necessitate that all students have access to a particular level or amount of technology, or workplace programs, where employers similarly find it beneficial to have all employees equipped with at least a basic computer.

In developing countries, the issue of affordability of computers is of course more critical. For example, twenty of the thirty-five least developed countries maintain statistics on personal computers per capita. Among these twenty countries, the median level of personal computers per 1,000 people is only 2 (UNDP 2000). In India, which is at the lower end of the next tier of countries (medium development), the level is 3 computers per 1,000 people. Even in China, where economic growth has created one of the largest information technology markets in the world, the level is only 9 computers per 1,000 people. These statistics are not surprising, given the low GNP per capita: $780 in China, $450 in India, and $410 in the low-income countries as a whole (*World Development Report 2000/01*).[4]

Innovator's Dilemma

The cost of computing power has fallen so dramatically that in theory even the poor should be able to afford computers. However, there is an important socioeconomic paradox at work here that hinders the ability of the unfettered market to function on behalf of the poor. This paradox is known widely as the innovator's dilemma (Christensen 1997). Simply put, companies and entire industries, as they become more successful, are driven to seek higher profit margins in order to survive. Once established markets become saturated, these higher profits are attained by means of adding new value to products, targeting the high-margin but narrow upper end of the market. In contrast, there is little incentive to create "disruptive technologies" (Christensen, Craig, and Hart 2001) that might bring entirely new product lines to as yet untapped but risky and low-margin mass markets.

One well-known example of adding new value to existing products was the path IBM took in creating more expensive and powerful mainframe computers for businesses while failing to see the potential of a brand new market for lower-cost, lower-performance personal computers. Similarly, today, the same computer companies that overtook IBM now devote their efforts toward making personal computers ever more powerful to achieve higher margins from U.S. business purchasers rather than initiating entirely new lines of inexpensive computers that would meet the needs of the poor in developing countries. This high-end emphasis is evident not only in hardware development but in software as well, with the dominant Microsoft operating system and office suite costing more than the annual income of most people in the developing world.

There are several compelling reasons that industries pursue the path of upper-market share rather than developing disruptive technologies (Christensen 1997). First, companies depend on customers and investors for financial resources, and it is the highest-performing companies in developed countries that are the largest and richest customers and investors. Investments and purchases by these companies strongly influence research and development within technology-oriented companies. Second, small markets don't solve the growth needs of large companies, and any brand new market is inevitably small to begin with (even though it may have the potential to become huge over time). Third, markets that don't yet exist cannot be analyzed or forecast, so planning new product lines for new markets is always a high-risk enterprise; note, for example, the ongoing financial woes of the Internet-based company Amazon.com, which first broke open the online shopping market. Fourth, products devoted toward the lower end of the market inevitably have lower profit margins. Hence, companies and industries spend far greater time, effort, and resources developing extravagant and expensive products whose full capacity is needed only by a small proportion of the world's population (think of anything from leather-interior sport utility vehicles to the bloated Microsoft Office software suite) rather than creating cheap, practical, mass-produced products that can serve the needs of hundreds of millions of people around the world.

The huge gap between the most developed countries and the rest of the world—not only in terms of income level but also in terms of physical, social, and legal infrastructures—exacerbates this dilemma. Once markets are saturated in developed countries, companies and industries do often seek to expand to the developing world, but they inevitably target the relatively small upper middle class in these countries. It is far safer—and, in the short run, more profitable—for Hewlett-Packard to sell its existing personal computers to businesses in Shanghai or Beijing rather than to try to develop an entirely new type of computer for China's, and the world's, poor.

What about companies in developing countries themselves? The largest and most successful ones are locked into the same dynamics as big companies in the developed world, pursuing larger profits through targeting the higher ends of the market in their countries and abroad. Smaller companies with social or economic motivations for developing products for the rural poor almost always lack the kind of venture capital that is required for major new technological product development. However, as Christensen, Craig, and Hart (2001) point out, the governments of large developing countries have greater resources than even the largest corporations. They are well able to lend an important hand in promoting the development of disruptive technologies that might serve the needs of their poor. Indeed, this is beginning to happen in some developing countries in the area of computer technologies.

Brazil's People's Computer   One effort under way with governmental monetary and policy support is the development of a *computador popular* (people's computer) in Brazil.[5] Initiated by the national government, a Laboratory of Universal Access was established at the Federal University of Minas Gerais to build a low-cost, low-maintenance personal computer for the Brazilian masses. A team of professors, researchers, and graduate students has developed a prototype of a machine that resembles an ordinary personal computer with a 500 MHz processor, 64 MB of main memory, a color monitor, speakers, a keyboard and mouse, a 56 KBps modem, and extra modules so that a printer or disk drives can be added later. The main innovation concerning the device is a 16–32 MB flash chip that substitutes for a hard drive. This inexpensive nonvolatile chip has no moving parts, maintains its memory during power outages, and does not break down. The chip will come pre-installed with a free software package, including the Linux operating system and a Linux-based office suite and Internet browser. A portion of the flash chip's memory will be available for user data storage. Additional storage can be handled by local or remote servers.

Brazil's National Bank for Economic and Social Development is providing capital to manufacturers of the *computador popular*, and the government has already committed to purchasing 60,000 of the computers for use in schools. The people's computer will cost about $250 USD. Brazil's federal savings bank is establishing a special credit program for purchasers, who will pay approximately $10 USD per month for the computer over a period of three years.

India's Simputer   A second, more ambitious plan is advancing in India for a Simputer,[6] or simple, inexpensive, mobile computer. The Simputer project was first conceived during the planning of an international seminar on information technology for developing countries held in Bangalore in October 1998.

One outcome of this conference was the formation of the Simputer trust by seven prominent Indian computer scientists in order to carry out research and development in the manufacture of low-cost computational devices, minicomputers, and computer- and Internet-related applications for rural, semirural, and lower-income people; to manufacture prototypes, simulations, and mock-ups of such devices and applications; and to encourage more widespread use of these technologies among the poorer sections of the country. The group launched a prototype of the

Simputer in April 2001 and is now seeking to license it to manufacturers for mass production.

Unlike the Brazilian people's computer, which is essentially a simplified version of a desktop computer, the Indian Simputer represents a radically new handheld computing device, with its hardware and functions planned in accord with the needs of India's working people. The machine looks at first glance like a slightly oversized Palm Pilot but is actually a pocket computer with 32 MB of RAM and a Linux-based operating system. Limited internal storage is available through flash memory, and a plug-in smart card allows the Simputer to connect wirelessly to a local server. The device is powered by regular AAA batteries, rechargeable batteries, or via a cord to a power outlet. The primary output is by image or sound, with an extensive amount of audio available in the form of text-to-speech or prerecorded audio snippets. The primary form of input is through touch, with items selected by touching them on the screen, but users can also tap in characters on a "soft keyboard" (that can be brought up on the screen, Palm-style) or trace the characters in a manner similar to Palm's graffiti system. Input and output are available in several of India's main languages. Thus neither knowledge of English nor literacy in any language is a prerequisite for using the Simputer. Finally, an especially rugged case offers protection against harsh environments.

The Simputer is designed to be sold for $200 USD, which is still beyond the reach of most of India's poor. However, it is also specifically designed to be shared by a collective of users. For example, the smart cards developed for use with the Simputer can contain personalization information required to log on to a community server that maintains personalized data about the user. A rural community could thus own several Simputer devices and rent them out for use to individuals who own the inexpensive smart cards. In this manner, a farming collective, artisan group, or rural telecenter could purchase Simputers for cooperative or public use.

The major projected uses of the Simputer include communication (e.g., checking if somebody in the next village is available for a meeting); transportation arrangements (e.g., reserving train tickets in the nearest city); microbanking (e.g., making utility payments and getting on-the-spot electronic receipts); rural health (e.g., collection and sharing of village health statistics and information); e-governance (e.g., accessing land records); and literacy development (based on features such as high-resolution text displays combined with audio files and text-to-speech in local languages).

Also innovative is the licensing plan proposed for Simputer products. Drawing inspiration from the open source software movement (e.g., Linux), the Simputer Trust plans to release the product through what is being called open hardware licensing. Manufacturers will be given the nonexclusive rights to manufacture the Simputer for a nominal fee and, if they wish, to modify and extend the Simputer's specifications. After a one-year window of development and marketing opportunity, the hardware design that they create will then revert to the public domain. The Simputer Trust members hope that this arrangement provides sufficient incentive to innovate while keeping the Simputer as a publicly accessible tool that can be further developed by many while still remaining affordable.

The Simputer has generated a great deal of anticipatory excitement in both India and other developing countries. It has also been met with some skepticism, especially in the West. As one U.S. high-tech entrepreneur wrote in a spirited discussion on Slashdot (the main Internet forum for computer discussions, with 30 million postings a month), "Why does the 'poor illiterate farmer' out in the fields need a computer? Just because you can mass-produce an inexpensive computer for the masses, doesn't necessarily mean that everyone in the masses actually needs one. Or wants one for that matter."[7] A reply by an Indian correspondent tried to convey the more complex reality of India and the role that information and communication technologies could play for rural development:

In India the development is very uneven. So, there are a few states which are very poor and most which are OK and some really well off. What's happening out there currently is that the OK and the well off states actually have started giving computer access to people in villages. As to what they use it for—an example from actual usage—a soybean farmer finds out the price of soybeans in Chicago, because the price in Chicago affects the price in India in a few months, so, he can decide how much to sow. Also, when he is ready to sell his soybeans, he finds out which market gives him the best price and rents a truck to sell there. A widow is not getting a pension for her husband because of bureaucracy, she goes to the village computer and pays about 10 cents to send an e-mail to a high up official. He responds and she starts getting the money.[8]

Whether the Simputer will ever fulfill the ambitious visions set for it is yet to be seen, and indeed, there remain many challenges to move the Simputer from prototype to mass production, let alone to integrate it fully into the kind of rural development uses suggested in the previous posting. Nevertheless, a significant number of these types of rural computer uses are already occurring in India (see chapters 4 and 6) and if the Simputer is successfully manufactured and distributed, it will likely contribute to the expansion of these extant practices. In the meantime, the very existence of the Simputer Trust and project has focused the energy of many leading engineers in India on trying to develop a device and infrastructure that fit the social, economic, and linguistic context of Indian society.

Telecommunications

The second arena for potential intervention, both in developed and developing countries, is telecommunications. Two issues of access provision are involved: the first is extending the telecommunications infrastructure throughout the country, and the second is making telecommunications and Internet accounts affordable to individuals. Again, for purposes of discussion, I address these issues separately for developed and developing countries.

Developed countries have, for the most part, solved the problem of extending basic telephone infrastructure throughout the country. In most developed countries, however, infrastructure inequality still exists in the area of broadband access. High-speed Internet access through cable modem and DSL lines is generally most available in urban and suburban areas, with limited availability in rural communities. It will be important in the future to monitor the issue of broadband infrastructure and see if urban-rural gaps are sufficiently met by market forces or if a persistent gap requires some kind of policy intervention. It is also worth noting that broadband access is expanding most rapidly where competitive services exist (e.g., cable modems vs. DSL providers, or multiple DSL providers with their own infrastructures; see OECD 2001). A highly competitive market helps account for the rapid expansion of broadband in South Korea, for example. In contrast, where there have been no competitive alternatives, the rollout of new broadband services has been much slower.

In the meantime, a more immediate problem for developed countries is the affordability of regular Internet access via telephone modems. The obstacle here is not the cost of the monthly Internet account, which is in most cases relatively inexpensive and often can be had for free, but rather the cost of metered telephone use charges. The OECD has calculated the cost of a monthly "Internet Access Basket," consisting of the cost of a phone line, forty hours of peak time phone usage, and an Internet account, in the thirty-one OECD countries. The variation in cost is great, from a high of $175 USD in the Czech Republic to only $25 USD in the United States. Unsurprisingly, those countries with greatest competition among telecommunication services tend to have the lower service rates, in large part because of the existence of unmetered telephone calls (or, alternatively, low per-minute costs).

As mentioned previously, developing countries face much more serious challenges in the area of telecommunications than do developed countries. These include extending existing wired telecommunications infrastructure and services throughout the country, making these wired phone services more affordable to individual users, and potentially leapfrogging to new forms of wireless connectivity. As is the case with telecommunications in developed countries, a key factor in addressing these challenges is telecommunications competition.

Scott Wallsten (2001) studied telecommunications performance in thirty African and Latin American countries. The study examines the impact of three factors—privatization, competition, and regulation—on telecommunications performance. Wallsten found that of the three factors, competition was the most effective agent of improved performance; greater competition was significantly associated with increases in the per capita number of telephone mainlines, the number of payphones, and general connection capacity, and a decrease in the price of a local phone call. Interestingly, privatization in and of itself had little benefit for users. Indeed, its only positive correlation was with an increase in the number of payphones available to consumers, and it was actually associated with a decrease in general connection capacity. However, when privatization was combined with the existence of a separate regulator, it had several positive associations, including increases in connection capacity, labor efficiency (measured by number of mainlines per telecommunications employee), and number of mainlines per capita. Wallsten's research refutes the view that privatization alone will bring improved communications service. His data show instead that competition, not mere privatization, is the real goal in any efforts to promote telecommunications diffusion and that competition is accomplished best through a combination of privatization plus effective regulatory reform.

However, increased competition alone will not resolve the telecommunications problems in the least developed countries. To do so will require leapfrogging via disruptive wireless technologies. Leapfrogging in a developmental context has been defined as the use of new technologies to either speed through or jump over stages of development (Singh 1999). It also can refer to skipping over a technological frontier or product cycle. In the context of telecommunications and the Internet, it can mean skipping to a more technically advanced telecommunications system in order to overcome the lack of land lines and thus assist a country in reaching a more advanced communications infrastructure. Ashkok Jhunjhunwala (2000), director of the Telecommunications and Computer Networks Group (TeNeT) at the Indian Institute of Technology, Madras, has put forward a convincing analysis of why telecommunications leapfrogging is necessary in a country like India. According to Jhunjhunwala, a telecommunications company in the United States, receiving an average of $360 per year ($30 per month) from a typical household account, can easily afford to invest $1,000 for each new line it installs. Therefore, the focus of research and development in the United States has been not so much on reducing per line cost but rather on providing a larger basket of services at a higher cost.

However, in a country such as India, assuming that a household is able to spend 7% of its income on telecommunications, only 1.6% of households would be able to afford a telephone at the cost of $350 USD per year (table 3.5). With current technology it actually costs about $800 for a telecommunications company to install each telephone line in India. With a need for a 35% return per year (based on 15% finance charges,

Table 3.5
Percentage of Indian Households That Can Potentially Afford Telephone Service, 2000

Annual Income	Households (%)	Affordable Annual Expenditure on Communications (USD)	Cost Per Line Investment Supported (USD)	Total Households Reached (%)
>$5,000	1.6	>$350	$1,000	1.6
$2,500–$5,000	6.3	$175–$350	$500–$1,000	7.9
$1,000–$2,500	23.3	$70–$175	$200–$500	31.2
$500–$1,000	31.8	$35–$70	$100–$200	63.0
<$500	37.0	<$35	<$100	—

Source: Adapted from Jhunjhunwala (2000).

10% depreciation, and 10% operation and maintenance), Indian telecommunications firms would actually need a monthly income of $280 USD per year per line, which is affordable to only some 3% of Indian households (corresponding closely with the current teledensity rate in India). If the cost per line installation could be brought down from $800 to $200 per month, then telephone service would be affordable by some 30% of Indian households, multiplying home phone access tenfold. However, an installation price drop of that scale will never be met through increased competition alone, since the wiring of individual homes is an expensive proposition anywhere in the world, and especially so in India.

In response to this conundrum, TeNeT has developed a wireless local loop system (Jhunjhunwala 2000) that can skip over the cost of wiring each household—the so-called last mile cost that is the plague of telecommunications development all around the world and that accounts for some two-thirds of the per line cost of installing phone lines. The wireless loop system can provide simultaneous Internet access and voice communication telephone services using satellite-based connections to 200–2,000 households within a 10–25 km radius at a current cost of roughly $375 USD per line per year in urban areas and $425 USD per line per year in urban areas. Jhunjhunwala believes this price can be driven down still more—to approximately $250 USD per new line—by creative use of the large informal labor sector in India. For example, cable television services already make use of self-employed people to visit individual homes, sell and install dishes that connect to cable television, and revisit cable-connected homes regularly to collect monthly charges (all at an operation cost of less than one-third of that which would be incurred if the organized corporate sector took on the same tasks). This strategy has provided livelihood to large numbers of self-employed people while driving down the cost of cable television to less than $4 USD per month in India, thereby contributing directly to its extremely rapid growth rate (from 0 in 1992 to 40 million households in 2000) (Jhunjhunwala 2000). It is reasonable to assume India's large numbers of self-employed people could be similarly effective in providing wireless local loop telecommunications connectivity, especially because the technology connects via a dish and thus requires less expertise than regular phone line connections (or even cable television).[9]

The wireless local loop system has been licensed for production to several manufacturers in India, Tunisia, and Singapore, and is already being deployed in Argentina, Brazil, Madagascar, Nigeria, Tunisia, Kenya, Angola, Yemen, Fiji, and Iran. Pilot use in India includes the built-up business district of Delhi (thus providing the "last mile" in an area that already had main phone lines) and a rural network of sixty-five villages in Andhra Pradesh, which previously lacked telephone and Internet access but now has both.

There are still many technical obstacles to further driving down the cost of wireless local loop systems, improving their capacity, and producing and installing them widely. More important, however, there are also large social, economic, and political barriers to developing and implementing this kind of disruptive technology on a mass scale. Telephone networks in India, as in other countries, are controlled by a handful of large companies (including the national government company and a few large international companies), none of which has put much emphasis on serving rural areas where population density is low and establishing new lines is expensive. However, neither do these companies like to see new competitors come onto the scene that could eventually challenge their current market dominance. At present, market control by a few major players is ensured by the prohibitive cost of new licenses to provide telephone lines, with no exceptions made for smaller companies that primarily or exclusively serve rural customers. Large telecommunications companies also have huge amounts of capital, which they use to provide all sorts of generous vendor financing. In contrast, local franchisees in India have little capital, so they depend on up-front vendor financing to get started, thus providing another important advantage to the major companies.

The major telecommunications companies currently operating in India are developing their own wireless networks, but these are designed to add value to existing telecommunications services and are aimed more or less exclusively at the already connected urban population, through, for example, expensive mobile telephone service as an adjunct to home or business connections. There is no indication that these companies have yet devoted significant research and development effort to the rural market. The Indian government, which itself owns the largest telecommunications company in the country (and also has amicable relations with the other large telecommunications companies), has yet to back the wireless local loop system. Ironically, at this point the government may actually be encouraging obstacles, for example by excluding smaller and newer telecommunications companies from bidding for large governmental contracts.[10]

The case of telecommunications development in India is a compelling example of the innovator's dilemma. Adding to this dilemma is the fact that the cost of installing new telephone lines, unlike that of producing computers, has not fallen significantly in recent years. This is a clear example where a disruptive technology is required to meet the needs of the hundreds of millions of people around the world who cannot afford telephone service. As I have already argued, the obstacles to diffusing such technology are not so much technological but social, economic, and political. Market-oriented systems reward and protect those that are developing products and services for the already existing market, especially the business sector and middle- and upper-middle-class consumers. In order to counteract this tendency, other kinds of support and incentives, most likely from governments, will be required to nurture technologies that meet the needs of low-income populations who only represent potential rather than existing markets.

Public Access Centers

Market expansion and decreased price of computers and telecommunications are essential to increasing Internet access around the world. At the same time, however, it will be decades before nearly every household in developed countries has Internet access, and much longer than that before universal home Internet service is reached in developing countries. It is thus necessary to enhance the provision of Internet connectivity through the establishment of public access sites.

Public access centers go by many names. The most common names used internationally are cyber café, telecenter, and community technology center. Other names have sprung up in particular countries or regions, such as computer kiosk (India) or *cabína publica* (Peru). Regardless of name, public access centers have many common features. They offer opportunities to use computers and the Internet without home ownership of a computer or a telephone line. In many places, the quality of computers and the speed of Internet access are higher than that ordinarily available in most homes. Many centers also offer forms of hands-on guidance, support, or training to users.

Public access centers also differ significantly along a number of axes (table 3.6). The first axis concerns management type. Centers can be run by commercial institutions, governments, schools, universities, and nongovernmental organizations. In some cases they are individual centers, in other cases, individually operated but part of a network or franchise, and in still others, multiple individual centers under the same administration or ownership.

A second axis pertains to location. Centers can operate in dedicated buildings, cafés, schools, universities, government offices, libraries, and community organizations and centers. They can be found in both urban and rural areas.

The third axis of difference relates to function. All centers offer some type of individual computer or Internet use, but some also offer training, community content production (e.g., flyer production for announcing forthcoming events), broader public services (e.g., mail service or utilities payment), or broader private services (e.g., banking activities or equipment rental). Centers also differ as to whether their main goal is to provide individual access to computers and the Internet, to train people in computer skills, or to assist community development.

Table 3.6
Variables in Internet Public Access Centers

Management		Location		Function	
Type	Administration	Site	Community	Service	Purpose
Commercial	Individual	Café or restaurant	Urban	Individual computer and Internet use	Individual access
Municipal	Multiple	Dedicated telecenter	Rural	Training	Education
School	Franchise	Public library		Content production	Community development
University	Network	School		Broader public services (e.g., ordinary mail, utilities payment)	
Nongovernmental organization		University		Broader private services (e.g., banking, office or equipment rental)	
		Community center			

Centers whose main function is to provide education or to assist with community development are discussed in more detail in chapters 5 and 6. For now, I focus on centers whose main purpose is to increase physical access to computers and the Internet. As in other issues related to connectivity, important differences exist between centers located in developed and developing countries. In developed countries, cyber cafés— coffee shops that include computers and Internet connections in their offerings—originally served primarily university students and young urban professionals. The emphasis was often as much on the café as it was on the cyber. Subsequently, and with Internet access becoming more common in developed societies, public access centers have in the past five years migrated toward serving the poor. Today, one is likely to find commercial cyber cafés with a clientele of immigrants writing home rather than of students keeping in touch with friends (Colker 2001). In addition to the rise of cyber cafés, long-standing public institutions, especially public libraries, are seeking to provide Internet access to the general public. In general, these public institutions target users who cannot otherwise afford to use the Internet. Community technology centers (CTCs), many of which see their goals as explicitly focused on community development rather than merely on individual access (see chapter 6), have also embraced the need to provide underserved communities and social groups with free or cheap access to telecommunications.

In developing countries, on the other hand, only a small percentage of the population can afford home use of computers and the Internet. Thus, public access centers in many locations are not an auxiliary method of accessing the Internet but the main method. A great deal of attention has gone into their design, implementation, and evaluation, especially in Latin America and the Caribbean, where many countries have sufficient resources to provide some telecommunications infrastructure but are not yet wealthy enough to allow for majority home access to computers or the Internet. Public access centers, generically referred to as *telecentros* (telecenters), are relatively widespread in Latin America and have been studied and discussed extensively (e.g., Gómez, Hunt, and Lamoureux 1999; Hunt 2001). A study by Proenza, Bastidas-Buch, and Montero (2001), based on direct observations in Peru, Panama, El Salvador, Guatemala, Brazil, and Chile with additional data collected from the remaining Latin American countries, is especially insightful in relation to telecenters. Their conclusions are as follows:

• A telecenter can be a powerful instrument for development but, to be effective, it must be part of a comprehensive economic and rural development strategy that also includes other institutional reforms to broaden the work opportunities and social and economic participation of traditionally excluded sectors of the population.

• Market incentives are important for the success of telecenters, but are not sufficient; particularly in remote and sparsely populated areas, government subsidies are required to make public access centers viable. This is already being done in some countries by means of government subsidies for public telephone services that help defray Internet connection costs for telecenters. The benefits of these investments should be maximized by extending them to Internet service as well as rural telephony.

• Telecenters can provide an important complement to formal education reform by providing support for students and teachers in relation to after-school computer and Internet use and by increasing Internet access for teachers, parents, recent graduates, and the community at large outside school hours. The educational impact of telecenters will be maximized if governments work simultaneously to strengthen the formal education system so that it routinely incorporates the effective use of new technologies.

• The impact of telecenters is multiplied if governments also give priority to the development of online content and public services aimed at meeting the economic and social needs of low-income populations, including portals and Web sites that use simple language and that broaden labor and self-employment opportunities.

• One of the main benefits of telecenters can be the promotion of virtual activism with real effects as a means of empowering low-income populations to be able to address their own problems constructively and effectively. This can be facilitated by the development of low-cost software tools for interaction and collaboration over the Internet and by telecenter administrators' supporting forums and projects that promote citizen interaction and social coordination.

• Successful telecenters must target a low-income population; retain strong commitment to self-sustainability and adopt a business model consistent with that commitment; and be backed by a leader who has a strong personal commitment, is willing to contribute his or her capital and time, is knowledgeable about the initiative's technical and financial requirements, and is willing to address the community's needs and investments.

Several of these points—such as education, community empowerment, and content development—are discussed at length later. I now turn specifically to the issue of content.

4

Digital Resources: Content and Language

Computers and the Internet are not much use without content and applications that serve people's needs. With the surge of material published in recent years on the World Wide Web—and millions of more Web pages added every month—it might seem that any shortage of online information and content has been long overcome. And from the point of view of a middle-class English-speaking American, that may well be the case. However, for those who live in different sociocultural environments or speak different languages, the situation is often very different. As this chapter argues, the massive amount of digital content being created on the Internet does not necessarily meet the needs of diverse communities around the world, and this has important consequences for issues of social inclusion. Fortunately, some excellent models exist of ways to develop relevant local content through active participation of diverse groups, and these models will be highlighted.

Global Web Content Production

It is impossible to determine the exact number of Web pages in the world, though current estimates top one billion.[1] It is somewhat easier, though still not exact, to count the number of Internet host domain names, such as www.harvard.edu, www.nytimes.com, or www.greenpeace.org, each of which, of course, can contain any number of pages within them. The number of Internet host domains rose nearly a hundredfold from 1993 to 2001 and now tops one hundred million.[2]

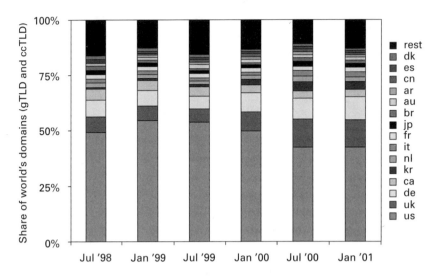

Figure 4.1
Share of world's Internet domains, by country, 1998–2001.
Source: Zook (2001a). Copyright 2001 by Matthew Zook. Used with
permission.

Matthew Zook of the University of California, Berkeley, has conducted extensive analyses of the concentration of domain names by city
and country around the world (e.g., Zook 2001c). According to his latest
figures, some 65% of the domains in the world are located in the United
States, the United Kingdom, and Germany—a figure that remained fairly
steady from 1998 to 2001 (figure 4.1). This signals a great disparity in
terms of number of domain names per person, even among the wealthy
countries themselves (figure 4.2), let alone between the developed and
developing countries. Within individual countries, the most domain
names are located on servers in major cities. Thus Internet content is
overwhelmingly concentrated in the major cities of the United States and
Europe, with only a few other key Internet server sites located in East
Asia, the Middle East, and Latin America, as illustrated by Zook's map
of global domain names (figure 4.3). There is also great disparity in
regard to representation of languages online; this is a pressing issue and
is the focus of the second half of this chapter.

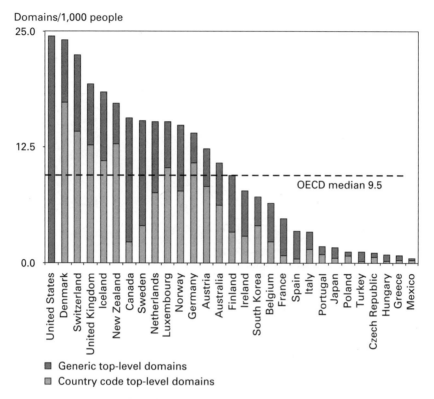

Domains/1,000 people

Figure 4.2
Internet domains per 1,000 people in OECD countries, January 2000.
Source: Zook (2001b). Used with permission from Matthew Zook.

Content and International Development

The geographic imbalance of Internet content production suggests that the content needs of diverse communities are not being met. For example, small-scale farmers and agricultural laborers in rural areas of Africa, Latin America, and Asia have little use for the types of material currently available on the Internet in their languages, and these rural areas of developing countries are almost completely uninvolved in production of Internet content. As a representative of M. S. Swaminathan Research Foundation told me, "The villagers in Kannivadi [in southern India] are not interested in what's going on in the White House or even in Chennai

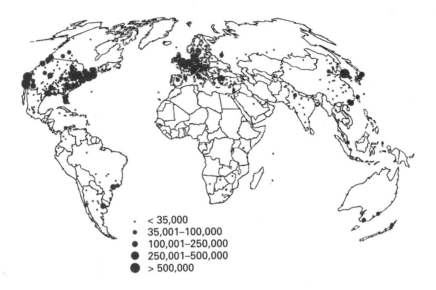

Figure 4.3
Total number of top-level domains (.com, .net, .org, country codes), by city, July 2000.
Source: Map based on methodology described in Zook (2001c). Used with permission from Matthew Zook. See also Zook's Internet Geography Project, http://www.zooknic.com/.

[the state capital]; they are interested in the price of rice in the local market." Governments, nongovernmental organizations, and community groups seeking to use the Internet for social development thus have to pay serious attention to the question of creating new digital content.

Some of the content and application areas that have been targeted as important for international development include the following.

Economic Development Information
In countries such as India, the population is made up mostly of small farmers raising a few crops each year on tiny plots of land. These farmers can benefit greatly from greater access to key information. For example, small-scale farmers often suffer financially from not knowing current crop prices in various markets in nearby cities. In response to this dilemma, several Internet projects for rural development in India collect crop prices and post this information as part of their intranets. For

example, in the Gyandoot Internet kiosks in the Dhar district of India, a farmer can come to a local kiosk and pay the equivalent of $0.10 USD to receive the market prices on that day for a particular crop at several local, regional, and national sales points. The farmer can then make a better-informed decision about whether to harvest the crop soon or let it continue growing and, when harvested, where to sell it for the best price.

Other types of information of value to small-scale farmers include data on soil testing, crop management, crop rotation, local crop varieties, and composting. Local projects in India, such as the M. S. Swaminathan Foundation's village knowledge centers, have taken the initiative to gather this information, rewrite it in local languages, and make it available to small farmers through local and regional intranets. Farmers can drop into the village knowledge centers that are located in fourteen villages of Pondicherry and Tamil Nadu and request free information from the kiosk operators.

Health Care

Some of the most promising information and communication technology (ICT) applications for telecommunications development are in the area of health care. The village knowledge centers are also used in India to deliver health-related information to rural areas. This information includes topics such as prenatal care, postnatal care, child immunization, tropical diseases, and local and regional health care resources. Rural areas in India suffer from a lack of trained medical personnel. For this reason, community development groups are working to develop software applications that could assist health workers with early disease diagnosis and prevention efforts. For example, the George Foundation in Bangalore is developing a software application known as the Early Detections and Prevention System 2000 (EDPS2000). This application is designed to enable the early detection of diseases and nutritional deficiencies among the rural population. It is intended for use in primary health centers, which often are unable to deal with early detection of diseases because of a lack of multidisciplinary medical expertise and laboratory facilities. EDPS2000 consists of a database of disease characteristics and conditions and a symptom diagnosis program. The program prompts the user for a step-by-step description of symptoms, followed by subsequent diagnostic questions. The software is designed to identify whether blood, urine, stool, or other laboratory tests are required in response to the answers keyed in by a health care worker. The system also indicates whether further investigation by a physician is warranted. In addition, the software maintains an exhaustive database of patient medical history and treatment visits to the health clinic and allows gathering of statistics about diseases, deaths, vaccinations and inoculations, pregnancies, and contagious diseases, thus enabling improved rural health care management.

Internet-based applications services also serve rural health care needs. One of the largest and most important is HealthNet,[3] which is used by approximately 19,500 health care workers in more than 150 countries worldwide (Accenture et al. 2001). HealthNet offers e-mail connections through low-orbit satellites to medical personnel in various locations throughout Africa. Beyond that, it has created online content and applications that are of use to medical practitioners throughout the world. These include two weekly online newsletters that focus on health issues in developing countries (one newsletter deals with general health issues and the other reports on AIDS issues), links to disease-specific online information, discussion forums on topics related to medical and pharmaceutical issues in the developing world, and a GetWeb application that allows people to download this Internet-based information using basic e-mail functions. HealthNet has had its greatest impact in sub-Saharan Africa, where it has local affiliates in Eritria, Ethiopia, Ghana, Kenya, Sudan, Uganda, and Zimbabwe. HealthNet is used for long-distance consultation among doctors in different African countries, for the scheduling of medical appointments in rural areas, and for gathering medical data for clinical trials.

Education

Most people in rural areas need to travel outside their communities to further their education. Information about places to study and entrance requirements is often difficult to access in remote areas. The Internet offers a vehicle for gathering and making available information about schools, courses, fees, schedules, and sample examination questions.

One of the most popular uses of rural Internet telecenters in India has been to get information on examination results. This information is often lost or delayed in the mail, making it difficult for rural youth to make plans for following up on their education. The information is available on the Internet and can be downloaded free or for a small fee at village Internet centers.

In the long run, rural development organizations need to develop computer-based and online content and applications not just *about* educational opportunities but to be directly used in teaching and learning. The possibilities and limitations of online education are discussed in more detail in chapter 5.

Community Affairs and Culture

Low-income urban and rural groups in developing countries often lack resources to express and share their own community's culture. Since it is less expensive to produce on the Internet than via print, television, or radio, online publications can provide an excellent medium for sharing locally developed community content and can often contribute to minority language and culture preservation.

For example, São Paulo, Brazil, is one of the most socially, economically, and geographically divided cities in the world. The wealthy residents of the city, including many of Latin America's leading bankers, financiers, and media moguls, live a world apart from the working class residents of the city's infamous *favelas* (shantytowns). Newspaper reporters and television journalists also keep a distance from the *favelas* except to report on drug wars or murders. Slum dwellers thus lack cultural and news outlets that report on events directly concerning their own lives. A community coalition called Sampa.org has stepped into this gap and established a community news service that gathers and publishes online information about local affairs, community services, neighborhood news, and cultural events. Sampa.org has also established an MP3 (music file) server, so that local hip-hop bands can share their music with each other and with listeners. All of this, of course, involves the active participation of the community itself.

Online Content in the United States

Even the United States, which leads the world in Web site production, suffers from significant content gaps that affect underserved communities. An in-depth study of Internet content and diversity in the United States was carried out by the Children's Partnership (Lazarus and Mora 2000). The study combined discussions with user groups, interviews with community center directors, and the analysis of 1,000 Web sites linked to commercial and noncommercial portals to evaluate the extent to which currently available content meets the needs of diverse U.S. communities. They identified four main content-related barriers that affected large numbers of Americans.

Perhaps the greatest barrier was a lack of locally relevant information. According to the study, low-income users seek practical, relevant information that affects their daily lives, such as the following:

• *Education*. Adult high school degree programs, adult literacy programs, financial aid, homework assistance, telementoring

• *Family*. Low-cost child care, low-cost enrichment activities for children, public programs for families

• *Finances*. Public benefits news, consumer information, credit information

• *Government and advocacy*. Immigration assistance, legal services, tax filing support

• *Health*. Easy-to-understand health encyclopedias, local clinics, low-cost insurance resources

• *Housing*. Low-cost housing, low-cost utilities, neighborhood crime rates

• *Personal enrichment*. Foreign language newspapers and search engines, communities of interest for youth and adults

• *Vocational*. Low-cost career counseling programs, job training programs, job readiness programs; job listings

In some of these cases, the information may be located in print documents, but these documents are difficult to locate and obtain. In many other cases, general or partial information may exist online, but not information that is particularly suited to low-income communities. For example, online job services generally target the higher-end market rather than entry-level jobs. Similarly, most online housing services focus on the higher end of the rental market rather than on low-rent apartments.

A second need was for information at a basic literacy level. For example, there are a large number of tutorials online that cover different computer and Internet skills, such as the use of spreadsheets, Web page design, or photo editing, but these generally demand a high level of literacy. Materials tailored to limited-literacy populations are badly needed by community technology centers, which often present computer instruction for those with limited English-language or literacy skills.

A third need is for content for non-English speakers. While there is a large amount of information on the Web in languages such as Spanish, there is little public information in Spanish directed towards U.S. audiences. Users seek information related to governmental programs that affect them, for example, Medicare, taxes, voting.

Finally, more diverse cultural resources are desired. Although some local U.S. communities are starting to build a cultural presence online (e.g., HarlemLive), users still feel that far more needs to be done to develop Web sites that reflect diverse cultural heritages and practices (Lazarus and Mora 2000).

The Children's Partnership has recently developed a major new portal in an attempt to address the lack of online information for low-income Americans. The portal includes sections for planning personnel, such as administrators of community technology centers, and for the lay public on topics like health, housing, employment, education, and culture.[4]

Content for the Disabled

Both developed and developing countries require content that addresses the special needs of the disabled in format and subject matter. As to formatting, the best and most up-to-date source of information is the Web Accessibility Initiative, supported by governmental and nongovernmental bodies in the United States, Canada, and Europe. These Web sites describe how online content should be developed so that it can be accessed by the disabled.[5] A principal requirement is to provide a redundancy of output mechanisms, that is to ensure that all graphical content has a text equivalent (for the blind, who can then convert the text to speech); that all audio content has a text equivalent (for the deaf); and that animated graphics can be frozen (for those with attention deficit disorder or learning disabilities). It is also recommended that sites allow users to input via both keyboarding or pointing (e.g., a "submit" button can be designed to also accept the input of the letter *s*) and that sites use a clear, consistent, well-labeled format (to benefit all users, and especially those with disabilities). Accessibility criteria exist on a continuum, so it is difficult to determine the exact percentage of existing Web sites that are or are not accessible, but by any measure there is still a long way to go. For example, one report found that the Web sites of all nine U.S. presidential candidates in 2000 failed to fully meet even the first level of accessibility requirements for the disabled (Báthory-Kitz 1999).

Beyond formatting, it is also important to develop online content for the disabled. Some of the best work in this regard has been carried out in Europe, where several countries have Internet portals for the disabled, with information on rehabilitation programs, assistive technology, special education, workplace adaptations, legislation, and training (European Commission 2001a; 2001b).

Community Mobilization and Content Development

The successful development of online content demands the active participation of the communities that will make use of the materials. There are three principal ways that community participation is achieved: through needs assessment, database development, and content production.

Needs Assessment
The approach of Participatory Rural Appraisal (PRA) provides a model of how a community can be engaged in helping define and determine its own needs (Mukherjee 1993). PRA has been used in development projects throughout the world over the last two decades, evolving from a prior, similar approach known as Rapid Rural Appraisal (RRA) (Chambers 1992). PRA uses focus groups, interviews, door-to-door surveys, community meetings, and special participatory exercises to max-

imize a community's involvement in defining its own needs. Rural Internet projects in India, such as Gyandoot and the M. S. Swaminathan village knowledge centers, were launched through intensive PRA in local villages. This appraisal of needs helped determine what resources villagers already had access to and what resources they needed.

Database Development

One important area of online content is listings, maps, and databases of local community resources. The community itself ought to be centrally involved in gathering and mapping those data. For example, the Camfield Estates project in Massachusetts, which placed computers and Internet access in many apartments in a low-income housing area, brought together a team of community residents to survey residents' existing skills, capacities, and interests and to identify other local assets, such as businesses, churches, and child care facilities. This information became a key part of the Web portal that served the Camfield Estates community.[6]

Community members can also contribute to databases through online communication. The city of Muenster, Germany, for example, has published a database and interactive street map for mobility-impaired people, with detailed and easily accessible information on public institutions, recreation facilities, social services, doctors, and information bureaus with barrier-free or disability-compatible buildings and services (Neumann and Uhlenküken 2001). One interactive feature being developed will allow community members to contribute directly to the database so that they themselves can point out urban locations that need to be altered or physical barriers that need to be removed. This system of community contribution is designed to make the database more comprehensive and informative while actively involving the disabled as consultants and partners in the project.

Content Production

The third major area of community involvement is through specific content development. Teams of community residents can be trained to develop Web-based information about their community that focuses on news, current events, culture, or any other items of interest or concern. One example is São Paulo's Sampa.org project, discussed earlier, which involves teams of community residents in producing a local online news service. Another excellent example is HarlemLive in New York.[7] Harlem-Live is an Internet-based youth publication launched in 1996. It has a close relationship with the Playing2Win community technology center in Harlem, which hosts the publication on its Web site and provides office and production space for the publication's editorial team. Columbia University and a number of other local organizations provide additional support.

HarlemLive is a high-quality online publication, with general news reports, articles on community issues, arts and culture articles, photo galleries, a creative writing section, and a special women's section. The publication thus provides current, topical information by and for the Harlem community. Equally important, HarlemLive has trained several hundred Harlem young people as journalists, photographers, media administrators, Webmasters, and public speakers. The publication thus serves as a focal point for young people to develop and showcase their technical and communication skills while they address issues of concern to the community and create original content that helps give the community voice.

In summary, there are many types of online content of use to marginalized communities. Some of this content can be provided by outside agencies. But, for the most part, active involvement of the targeted populations—in defining their needs, collecting data, and authoring and publishing content—is usually required for success. This kind of approach, based on active community involvement, also helps guarantee the kinds of community training and mobilization necessary for long-term success.

Language

Language is one of the most complex and significant issues related to content and to broader issues of ICT and social inclusion. Language intersects with many other forms of social division related to nationality, economics, culture, education, and literacy. Language questions dramatically affect how diverse groups can access and publish information on the Web as well as the extent to which the Internet serves as a medium for expression of their cultural identities.

Language and Identity in the Age of Information

The critical role of language is situated in the broader social and economic transformation of recent decades. The information revolution, accompanied by the processes of international economic and media integration, has acted like a battering ram against traditional cornerstones of social authority and meaning. Throughout the world, shifts in economic and political power have weakened the role of the state, new forms of industrial organization have decreased the possibilities for long-term stable employment, and women's entry into the work force has shaken up the traditional patriarchal family (Castells 1997).

But every action brings a reaction. The last quarter-century has also witnessed a worldwide surge of movements of "collective identity" that "challenge globalization and cosmopolitanism" on behalf of people's control over their culture and their lives (Castells 1997, 2). These differ from earlier social movements, which in many parts of the world were based on struggles of organized workers. As Alain Touraine explains, "In a postindustrial society, in which cultural services have replaced material goods at the core of production, *it is the defense of the subject, in its personality and its culture, against the logic of apparatuses and markets, that replaces the idea of class struggle*" (quoted in Castells 2000b, 23, emphasis in original). Castells (1997) further explains the central role of identity in today's world:

> In a world of global flows of wealth, power, and images, the search for identity, collective or individual, ascribed or constructed, becomes the fundamental source of social meaning. This is not a new trend, since identity, and particularly religious and ethnic identity, have been at the roots of meaning since the dawn of human society. Yet identity is becoming the main, and sometimes the only, source of meaning in a historical period characterized by widespread destructuring of organizations, delegitimation of institutions, fading away of major social movements, and ephemeral cultural expressions. People increasingly organize their meaning not around what they do but on the basis of what they are. (3)

Within this contradictory mix of global networks and local identities, language plays a critical role. With other cornerstones of social authority, such as nation, family, and career, battered by the processes of globalization, language can become "the trench of cultural resistance, the last bastion of self-control, the refuge of identifiable meaning" (Castells 1997, 52). The struggle over bilingual education in the United States; the

Québécois, Basque, and Kosovar separatist movements; the battles over language and citizenship in post-Soviet countries; and language revitalization movements in Ireland (Gaelic), New Zealand (Maori), Morocco (Tamazight), and many other countries indicate the powerful role of language-based identity in today's world.

It is not surprising that language and dialect have assumed such a critical role in identity formation. The process of becoming a member of a community has always been realized in large measure by acquiring knowledge of the functions, social distribution, and interpretation of language (Ochs and Shieffelin 1984). In most of the world, the ability to speak two or more languages or dialects is a given, and language choice by minority groups becomes a symbol of ethnic relations as well as a means of communication (Heller 1982). In the current era, language signifies historical and social boundaries that are less arbitrary than territory and more discriminating (but less exclusive) than race or ethnicity. Language-as-identity also intersects well with the nature of subjectivity in today's world. Identity in the postmodern era has been found to be multiple, dynamic, and conflictual, based not on a permanent sense of self but rather on the choices that individuals make in different circumstances over time (Henriquez et al. 1984; Schecter, Sharken-Taboada, and Bayley 1996; Weedon 1987). Language, though deeply rooted in personal and social history, allows a greater flexibility than race and ethnicity, with people able to consciously or unconsciously express dual identities by the linguistic choices they make even in a single sentence (e.g., through switching or combining languages; see Blom and Gumperz 1972). By means of choices concerning language and dialect, people constantly make and remake who they are. For example, a Yugoslav becomes a Croatian, a Soviet becomes a Lithuanian, and a Canadian becomes a Québécois.

Yet, at the very time that linguistic diversity is becoming more critical than ever in people's lives and identities, a new communication medium has emerged that has been dominated by a single language: English. The dominance of English, not just on the Internet but also in many other international media and communications forums, has led to the rise of new concepts such as "global English." In order to fully appreciate the issues associated with ICT and social inclusion, it is necessary also to understand global English and how it has come to dominate digital telecommunications.

Global English

Although there have been many important international languages over time, including Latin, French, Russian, Chinese, Arabic, and Spanish, English is generally considered to be the first global language because of its current dominant role as a *lingua franca* in international communications. The rise of global English is the flip side of movements for local identity; it represents the need for an international medium of communication for global economic, political, and social exchange. According to information gathered by Crystal (1997), 85% of international organizations make use of English as at least one of their official languages, 85% of the world's film market is in English, and 90% of the published articles in leading journals of linguistics are in English.

Nevertheless, these statistics belie the fact that English is only spoken as a native language by a relatively small minority of people in the world. According to calculations, about 350 million people around the world speak English as a native language (Crystal 1997; Graddol 1997; 1999), representing some 6% of the world's population. This places English well behind Chinese in its number of native speakers, and not that far ahead of Spanish, Hindi, and Arabic, all of which may catch up or pass English in number of native speakers in the next sixty years (Graddol 1997). Another 350 million people are estimated to speak English as a second language, in countries such as India, Nigeria, the Philippines, and Singapore. There are also an estimated 700 million people who speak English as a foreign language, albeit with varying degrees of proficiency. Putting these numbers together, we see that three-quarters of the world's population knows almost no English, and even among the one-quarter who are said to speak it, the degree of competence varies markedly.

In many countries, unequal access to learning English overlaps with other social inequalities. Even though English is almost universally taught in secondary schools and universities, the majority of people in many developing countries never attend secondary school. Even those who do often face poorly trained teachers who do not speak English well themselves. Indeed, in many countries, the only reliable route to learning English is through expensive private education. With knowledge of English a requirement for access to many professions and university programs, English becomes one more barrier to equal opportunity for the poor. And even many people who speak English well may not be happy with the thought of its supplanting their own local language in as important a medium as the Internet.

English on the Internet
One of the first published studies of language on the Internet, and conducted in 1997, indicated that some 81% of international Web sites were in English ("Cyberspeech" 1997). At the time these results were made public, the dominance of English on the Internet caused great consternation around the world. Anatoly Voronov, director of a Russian internet service provider, voiced the sentiments of many when he said:

It is just incredible when I hear people talking about how open the Web is. It is the ultimate act of intellectual colonialism. The product comes from America so we either must adapt to English or stop using it. That is the right of any business. But if you are talking about a technology that is supposed to open the world to hundreds of millions of people you are joking. This just makes the world into new sorts of haves and have nots. (Cited in Crystal 1997, 68)

Within three years of this study, the percentage of English Web sites had fallen to 68% (Pastore 2000), still a sizable majority and well out of proportion to the number of English speakers in the world. A calculation of the ratio of Web pages to speakers of leading languages indicates that English speakers are still better represented and served on the Internet than speakers of other languages (table 4.1).

The question remains as to whether this drop from 81% to 68% represents the beginning of the end of English dominance online, or whether it marks a continuation of that dominance at an unacceptably high level. To better understand and interpret these figures, and to predict the likely trend in international communication online, it is necessary to distinguish between the short-term and long-term advantages that English has in the computing and Internet realms.

The short-term advantages were principally two: the Internet first arose in the United States and speakers of English were its designers—they thus wrote programs that relied on an English-language interface;

Table 4.1
Ratio of Speakers of a Language to Web Pages in That Language, 2001

Rank	Language	No. of Web Pages	No. of Speakers (thousands)	Speakers/ Web Page
1	English	214,250,996	322,000	1.5
2	Icelandic	136,788	250	1.8
3	Sweden	2,929,241	9,000	3.1
4	Danish	1,374,886	5,292	3.9
5	Norwegian	1,259,189	5,000	3.9
6	Finnish	1,198,956	6,000	5.0
7	German	18,069,744	98,000	5.4
8	Dutch	3,161,844	20,000	6.3
9	Estonian	173,265	1,100	6.4
10	Japanese	18,335,739	125,000	6.8
11	Italian	4,883,497	37,000	7.6
12	French	9,262,663	72,000	7.8
13	Catalan	443,301	4,353	9.8
14	Czech	991,075	12,000	12.1
15	Basque	36,321	588	16.2
16	Slovenian	134,454	2,218	16.5
17	Korean	4,046,530	75,000	18.5
18	Latvian	60,959	1,550	25.4
19	Russian	5,900,956	170,000	28.8
20	Hungarian	498,625	14,500	29.1
21	Portuguese	4,291,237	170,000	39.6
22	Greek	287,980	12,000	41.7
23	Spanish	7,573,064	332,000	43.8
24	Lithuanian	82,829	4,000	48.3
25	Polish	848,672	44,000	51.8
26	Hebrew	198,030	12,000	60.6
27	Chinese	12,113,803	885,000	73.1
28	Turkish	430,996	59,000	136.9
29	Bulgarian	51,336	9,000	175.3
30	Romanian	141,587	26,000	183.6
31	Arabic	127,565,000	202,000	1,583.5

Source: Adapted from Carvin (2001).

and the Internet, in its early iterations, functioned best in the ASCII code, which is very difficult to read and write with non-Roman alphabets.

These short-term reasons have already started to fade in significance and impact. As discussed earlier, Internet access is starting to reach saturation point in the United States but it is just taking off in many other countries around the world. As a critical mass of users gets online in a particular language, more people and businesses create Web sites in that language, and speakers of the language also have a greater number of potential partners for computer-mediated communication. This trend is also accelerated by the expansion of operating systems and Web page authoring software in non-Roman scripts, which allows people to communicate more easily in non-alphabetic languages such as Japanese, Chinese, and Hebrew. Because of these trends, the proportion of Web sites in English is expected to drop to 40% in the next decade (Graddol 1997).

However, even as English's short-term advantages decrease, it will still maintain a strong position over other languages on the Internet because of its long-term advantages. Principal among these is the historical fact that English was already the de facto global language at the time the Internet was created, and remains so today. The Internet, by enabling global communication, requires a global standard, and English's default advantage thus remains and is in fact strengthened. A mutually reinforcing cycle takes place, by which the existence of English as a global language motivates (or forces) people to use it on the Internet, and the expansion of the Internet (and online English communication) thus reinforces English's role as a global language. This cycle can occur even when more and more people are using the Internet in their own languages. They may use the Internet in their own language for local or regional communication, but they will continue to use the Internet in English for global communication.

This trend is illustrated in a study conducted by the Organization for Economic Cooperation and Development ("The Default Language" 1999). The OECD study found that while only 78% of the regular Internet sites surveyed were in English, some 91% of the sites on what are called secure servers were in English and 96% of the sites on secure servers in the .com domain were in English. This is significant because secure servers, especially in the .com domain, are most frequently used for e-commerce. This means that even as people increasingly use languages other than English for local communication, they can be expected to use English for many international transactions. Of course, this latter trend may not be permanent as more companies may localize their e-commerce in the languages of the consumers.

Furthermore, long-term advantages do not necessarily mean permanent advantages. It is entirely possible that a century from now English will no longer be the dominant language on the Internet (or on whatever has replaced the Internet), either because of the weakening of English as a global language (because of demographic or economic changes) or because of the development of improved machine translation techniques (thus allowing everyone to communicate in the local language). Machine translation already exists online, but it is of such poor quality (based on word-by-word translation) that it does not now mitigate the need for a *lingua franca*, nor is it predicted to be of sufficient quality to mitigate such a need for a long time. For the foreseeable future, then, a disproportionate percentage of the world's Web sites, especially those necessary for international exchange, will be in English, and that is an important factor limiting access to Internet content.

The role of English vis-à-vis other languages online can be illustrated through analyses from Egypt (where English is spoken as a foreign language), India (where it is spoken as a second language), and Hawai'i (where it is spoken as a first language, though in a diverse multilingual setting).

Language Online in Egypt

Egypt is an excellent example of a highly stratified country in which English plays a dual role. On the one hand, English helps connect Egypt to the world by facilitating international commerce, tourism, and exchange. On the other hand, unequal access to English within the country serves to heighten the nation's already substantial social and economic disparities.

Arabic is the official language of Egypt and virtually the entire population speaks a dialect of this, referred to as Egyptian Arabic. Those that can read and write also know Classical Arabic, the main written variety of the language. Other languages in the country include ancient Coptic (used in Coptic Christian church services), a variety of African languages spoken by refugees, and European languages used in business and tourism. The use of European languages in Egypt has a long history dating back to periods of French and British colonialism, and at the time the Egyptian elite often preferred to be educated in French or English rather than Arabic (Haeri 1997). Most recently, though, the use of English has far surpassed that of French and other foreign languages by Egypt's elite. English is used not only in communication with foreigners—for example, in international commerce—but also for internal communications in a number of privileged occupations, especially in the fields of information technology, engineering, medicine, dentistry, and sciences. It is not unusual for Egyptian professionals in these areas to hold their conferences or produce their publications in English, even if the intended audience is other Egyptians.

English is essential for participation in elite professions, yet it is spoken by only a small minority of the population, estimated at some 3% (Warschauer 2001b). English is a mandatory foreign language taught in all schools beginning in the fourth year of elementary school, but it is learned very poorly, if at all, because of huge class sizes, poorly trained teachers (many of whom themselves barely know the language), and the country's high dropout rate (with half the adult population completing less than five years of schooling) (Fergany 1998). The elite, many of whom are bilingual in English and Arabic, usually learn English in private schools (a large number of which offer English medium instruction), private tutoring, English-medium private universities in Egypt, and study abroad in England or the United States (see discussion in Schaub 2000).

Even though English is spoken by just a small percentage of the population in Egypt, it is a dominant language of the Internet in that country (Warschauer, Refaat, and Zohry 2000). Many Egyptian Web sites, including those targeted exclusively for use inside the country, are only in English (see, for example, figure 4.4, showing Otlob.com, a popular site for ordering food delivery from restaurants in Cairo and Alexandria). Some 70% of young professionals I surveyed use English

Figure 4.4
Ordering food in Egypt.
Source: http://otlob.com. Used with permission.

exclusively in formal e-mail communication (Warschauer, Refaat, and Zohry 2000). Arabic, when added in e-mail, is most often written in Roman characters and used principally for religious or highly emotive expressions.

The reasons for English's dominant role in Egyptian online communications are multiple. First, no single standard of Arabic-language computing has emerged yet, so Web producers are often forced to convert Arabic-language content into slow-loading images if they want to guarantee that their content can be read in Arabic. This lack of a common standard also discourages Arabic-language e-mail. In addition, the Internet first arose in Egypt in the very sectors that operate in English, such as the information technology industry and international businesses. Finally, the early adopters of the Internet in Egypt were mostly people who—owing to their schooling and work experience—write, compute, type, and keyboard better in English than they do in Arabic, and using English online thus comes naturally to them.

Some of these conditions are bound to change over time, especially with the emergence of common Arabic-language standards for computing. However, in the meantime, the 97% of the Egyptian people who do not know English are excluded from full access to Egyptian online content written in English, let alone international English-language Web sites.

Multilingual Computing in India

India is another country where English plays a stratifying, if also a unifying, role. People in India speak some 850 local and regional languages (Todd and Hancock 1987) of which 58 are taught in schools, 87 are used in newspapers, 71 in radio programs, and 15 in films ("Indian Languages" 2001). A total of 18 of these are considered official languages.

English is spoken by an elite throughout the country and is used in scientific, technological, and business communications. However, despite the national prominence of English in India and India's reputation as an English-speaking country, only about 5% of the people speak English (Crystal 1997). In contrast, nearly half the people of the country speak Hindi and almost another quarter speak dialects of Tamil, Bengali, Kannada, or Marathi ("India" 2001).

Indian-language computing thus is complicated by many factors, including the fact that each major Indian language has its own script. Not surprisingly, English has emerged as the dominant language of the information technology industry in India, which serves the industry's booming export business well. However, English is much less useful as a language of communication for national development purposes, especially for projects that target India's poor. In short, the potential of ICT for aiding rural development will not be reached without adequate Indian-language software and content.

For this reason, a number of Indian organizations are trying to develop software solutions to promote Indian language computing. One of the more promising is being produced by a group called Chennai Kavigal, which is collaborating with the Institute of Indian Technology in Madras to develop low-cost Indian-language software solutions for both Windows and Linux platforms. One of their products, a complete office suite with a word processor, spreadsheet, database, and presentation software, is being made available to development projects for only $6 USD. Other products include e-mail, paint, browser, and programming software. The products are being developed with special attention to the needs of Indian users. For example, all products have complete Indian-languages interfaces, including navigation menus. The e-mail software comes with separate password-protected folders, ideal for a situation in which many users share one machine. A compiler of the C and C++ computer languages enables programming to be done in English (essential for getting a job) but allows comments and error messages to be written in Indian languages, thus providing important support to limited-English speakers who are in the process of becoming software programmers. Finally, to encourage international communication between India's different regions, the word processing and communication software products are programmed to automatically convert from one language script to another. This is especially helpful for the many millions of people in India who can read and write one Indian language but speak others. For example, a writer of Tamil (but speaker of Hindi) can write a Hindi-language message in the Tamil script and have it automatically converted to Hindi script to be read by someone in a Hindi-speaking region of the country. These conversions can even be performed instantaneously using synchronous communication software so that one's own script appears on one's own screen while the other script appears instantly on the correspondent's screen.

These software solutions are helping rural development projects in India to develop Internet content in local languages. Much useful content in these projects is not necessarily original material but rather public information already available elsewhere on the Web that is carefully selected, translated into Indian languages, and presented in a simplified form in easy-to-navigate pages.

Language and Identity in Hawai'i

Multilingual software solutions are only the first step. To help promote technology for social inclusion, it is also necessary to use the software to develop and make available relevant content. This can be done by translating material and, more important, by developing original material in the minority language itself. One of the more interesting examples of minority language content development is from Hawai'i.

The situation in Hawai'i differs markedly from those in India and Egypt. In Hawai'i, almost everyone except some recent immigrants speaks English. Why then is Hawaiian-language computing and Internet content even necessary? To understand this, it is necessary to examine briefly from a sociopolitical angle the history of Hawaiian language use.

Hawaiian was the national language of the sovereign Hawaiian monarchy of the nineteenth century. At the end of that century, however, wealthy American landowners backed by the U.S. government overthrew the Hawaiian kingdom and forcefully incorporated Hawai'i as a U.S. territory. Laws forbidding the use of the Hawaiian language were passed and vigorously enforced through beating children who dared speak their native tongue in school (Wilson 1998). By the time Hawai'i became a U.S. state in 1959, Hawaiian was spoken by only a few thousand elders and the language was seriously endangered. The suppression of the Hawaiian language went hand-in-hand with the subjugation of the native Hawaiian people, whose numbers gradually diminished and who found themselves at the bottom of all social and economic indicators with little room to move in terms of rescuing their dying language.

Nonetheless, a strong Hawaiian resistance movement emerged in the 1970s to fight for Native Hawaiian rights, including revitalization of the Hawaiian language. As a result of this effort, the Hawaiian language was legalized and a number of Hawaiian immersion schools were established by the state government. New undergraduate and graduate programs in Hawaiian studies and Hawaiian language were launched in the state's public universities. Ever since, defense of the Hawaiian language has continued to be a central element for defense of the Hawaiian people. This is especially so because of the complex nature of Hawaiian identity. There are almost no "pure" Hawaiians left, although some 20% of the people in the islands have part-Hawaiian ancestry. In this context, many Hawaiians believe that revitalization of their language is critical to their survival as a people.

The Hawaiian language is no longer facing extermination; nevertheless, it still faces an uphill battle to be reestablished as a stable living language used regularly in daily life. There are no daily Hawaiian-language newspapers or full-time Hawaiian-language television stations. The handful of elders who learned to speak Hawaiian as a first language are dying. The Hawaiian immersion schools are trying to teach Hawaiian but are confronted by a lack of curricular content and a dispersion of its speakers in small pockets spread out over several islands.

It was within this context that Hawaiian educators from the University of Hawai'i launched Leokī (powerful voice), believed to be the first online bulletin board system that functioned completely in an indigenous language (Warschauer and Donaghy 1997). The provision of Hawaiian-language content and communication media was intended to boost the Hawaiian revitalization effort, and particularly the Hawaiian-language programs in the immersion schools. As Keiki Kawai'ae'a, the director of Hawaiian-language curriculum materials, explained,

Without changing the language and having the programs in Hawaiian, they wouldn't be able to have computer education *through* Hawaiian, which is really a major hook for kids in our program. They get the traditional content like science and math, and now they are able to utilize this *'ono* (really delicious) media called computers! Computer education is just so exciting for our children. In order for Hawaiian to feel like a real living language, like English, it needs to be seen, heard and utilized everywhere, and that includes the use of computers. (Quoted in Warschauer 1998, 147–148)

All the interfaces, menus, and content for Leokī are written in Hawaiian. The board also contains an extensive array of features (figure 4.5):

• *Leka Uila* (electronic mail). Each user has a private mailbox for sending and receiving mail to and from other users on Leokī as well as via the Internet.

• *Laina Kolekole* (chat line). An online chat area for real-time interaction. Users can also create their own private chat rooms.

• *Ha'ina Uluwale* (open forum). Public synchronous computer conferences for discussion, debate, and surveys.

• *Ku'i ka Lono* (newsline). Advertisements, announcements, and information about Hawaiian-language classes and important upcoming events.

• *Hale Kū'ai* (marketplace). Announcements and online order forms for the purchase of Hawaiian-language books and materials.

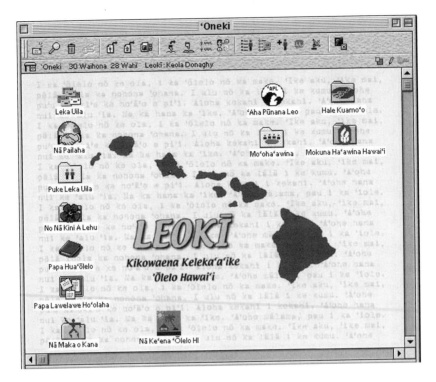

Figure 4.5
Hawaiian language bulletin board system.
Source: Kualono (2001). Used with permission from Kualono.

- *Papa Hua'ōlelo* (vocabulary list). Dissemination of Hawaiian words being coined by the Hawaiian-language lexical committee. Users can suggest new words and offer input on terms being considered or search vocabulary databases.

- *Nā Maka o Kana* (The Eyes of Kana). The current and all back issues of the *Nā Maka o Kana* newspaper, published by and for the Hawaiian immersion program.

- *Noi'i Nowelo* (Search for Knowledge). Shared resource area for old and new Hawaiian-language materials. Posting of stories, articles, and songs.

- *Nā Ke'ena 'Ōlelo Hawai'i* (Hawaiian language offices). An information section about the various agencies that provide educational support for Hawaiian studies coursework and Hawaiian-medium programs throughout the state.

Leokī has been installed on the computers in the Hawaiian immersion schools and preschools, Hawaiian-language departments of colleges and universities, and in Hawaiian-language support organizations. Most recently, a public version has been made available to any Hawaiian-language speaker.[8]

In addition, Hawaiian educators have actively mobilized the Hawaiian-language community to create content for Leokī and for the broader Internet. In 1997 I conducted an ethnographic study of a Hawaiian-language class at the University of Hawai'i that worked to develop a Web site in the Hawaiian language. The students in the course chose topics related to Hawaiian history and culture, including Hawaiian chanting and music, Hawaiian leaders, and the geography and nature of Hawai'i. The Web pages that they developed were highly sophisticated in language, content, and design, and the experience proved highly motivating for the students. As one of the students explained,

It's like a double advantage for us, we're learning how to use new tools, like new technology and new tools, at the same time we're doing it in Hawaiian language, and so we get to learn two things at once. . . . It looks almost as if it's a thing of the future for Hawaiian, because if you think about it, maybe there's [only] a few Hawaiian-language papers. But instead of maybe having a Hawaiian-language newspaper, you have something that might be just a little bit better, like the World Wide Web, it's like building things for all the kids who are now in immersion and even for us, someplace to go and get information, and so that's kind of neat what we're doing, we're doing research and then finding out all that we can about a topic and then actually putting it on the World Wide Web, and then having that be useful to somebody else in the future. (Warschauer 1999)

Nancy Hornberger (1997) once said that "Language revitalization is not about bringing a language back, it's about bringing it forward." For students in Hawai'i, indigenous-language content creation is bridging a very special divide, that between a recent past in which their language and culture faced persecution and near elimination, and a future in which their words, chants, songs, and stories will thrive with the assistance of new media.

Conclusion

Physical resources such as computers and connectivity mean little without sufficient digital content that is relevant to people and in the language of their communities. As many of these examples have shown, the most important content production is often done by people in the targeted communities themselves. This, in turn, demands literacy and education. It is to those issues that I turn now.

5

Human Resources: Literacy and Education

Literacy and education affect online access at both the macro and micro-levels. At the macrolevel, mass literacy and education serve to grease the wheels of economic development and thus create conditions for greater technologization of society. As Robison and Crenshaw (2000, 8) explain,

Mass education accelerates this structural [economic] transition by (1) "lubricating" the movement of workers between sectors by providing workers with the necessary cognitive skills and attitudes; and (2) encouraging rapid rural-to-urban migration as literate agricultural workers seek better lives in the city. As a simple extension of this logic, the demand for computers and on-line skills will be driven in part by the degree of education in the population.

Indeed, Robison and Crenshaw found in their large cross-national study that mass education correlated directly with high levels of societal Internet access.

Of course, mass education is not only a cause of economic development but also an effect. The complexity of the relationship is revealed by an important interactional relationship revealed in the Robison and Crenshaw study: The correlation between mass education and societal Internet access is highest in countries that have a large "tertiary" economic sector (based on consumer and producer services, typical of the postindustrial economy).

Education and literacy are also important at the individual microlevel, since reading, writing, and thinking skills remain crucial for being able to use the Internet. Education also helps determine *how* people use the Internet and what benefit they achieve from it. As the Internet becomes more widespread, it is highly likely that its use will be stratified, with some using it principally as an entertainment device and others using it to seek and create new knowledge.[1] The mere existence of the Internet will not create researchers or knowledge seekers out of those without the requisite background or skills.

How, then, can education and literacy best contribute to effective use of information and communication technology, and correspondingly, how can effective use of ICT contribute to education and literacy? To address these questions, I first examine how the Internet is affecting literacy and what new literacies are required for using the Internet. I then look at the broader question of education, analyzing how computers and the Internet are, or are not, transforming learning and teaching.

Technology and Literacy

All human activity is mediated by tools. What is significant about tools is not their own abstract properties but rather how they are incorporated into, and fundamentally alter, human activity. In other words, tools do not simply facilitate action that could have occurred without them, but rather, by being included in the process of behavior, alter the flow and structure of mental functions (Vygotsky 1981). The integration of tool, mental system, and human activity is illustrated nicely in Bateson's discussion of the blind man with a stick (1972, 459):

Suppose I am a blind man, and I use a stick. I go tap, tap, tap. Where do *I* start? Is my mental system bounded at the handle of the stick? Is it bounded by my skin? Does it start halfway up the stick? Does it start at the tip of the stick? But these are nonsense questions. The stick is a pathway along which transformations of difference are being transmitted. The way to delineate the system is to draw the limiting line in such a way that you do not cut any of these pathways in ways which leave things inexplicable. If what you are trying to explain is a given piece of behavior, such as the locomotion of the blind man, then, for this purpose, you will need the street, the stick, the man.

The tools of literacy include language itself as well as a wide variety of physical artifacts, such as the papyrus, codex, book, pencil, pen, paper, or typewriter. The development of each of these tools has had a profound effect on the practice of literacy.

Today, social, economic, and technological transformations are again aligned to bring about major changes in literacy practices. New types of computer- and Internet-based literacy practices are emerging that I have referred to previously as electronic literacies (Warschauer 1999; see also Shetzer and Warschauer 2000). Electronic literacies are not isolated from the types of literacy practiced with print but rather involve added layers that account for the new possibilities presented in the electronic medium of computers and the Internet (Buzato 2001; Selfe 1990). Electronic literacy is actually an umbrella term that encompasses several other generic literacies of the information era, including computer literacy, information literacy, multimedia literacy, and computer-mediated communication literacy. These new literacies stem in part from the new technological features of the computer but also from the broader social setting in which computers are used.

Computer Literacy

The term *computer literacy* emerged in the early 1980s together with the spread of the personal computer.[2] Within a decade, the term had become widely discredited among educators because it generally referred to only the most basic forms of computer operation, such as turning on a computer, opening a folder, and saving a file, and thus tended to justify a very limited view of computer-related education. While the criticisms of computer literacy as an end in itself are certainly merited, there does exist a fluency (or, alternatively, unfamiliarity and discomfort) with the physical and operational manipulation of a computer that profoundly affects people's productivity with it and that intersects with a range of social dimensions, such as age.

Consider, for example, the kinds of computer literacy involved in browsing or searching the Internet (which overlap with information literacy, see next section). Hargittai (2000b) conducted an extensive observational and interview study of a randomly selected Internet user population from the general public in New Jersey. She found that some people had little awareness of the Back button, a button ordinarily used in 30% of people's browsing activities (Tauscher and Greenberg 1997), which severely hindered their ability to navigate the Web. Many people had a difficult time entering valid search terms in large part because of spelling errors. Other users insisted on putting search terms together without any spaces (e.g., MichaelJordanbasketball) , mistakenly extrapolating from the fact that URLs cannot include spaces. And others rarely used search engines at all, relying solely on functions of their browsers or Internet service providers.

An example of the value of computer literacy is offered by the renowned Colombian author, Gabriel García Márquez, who has described how his own productivity has multiplied greatly through his writing on computer (Day and Miller 1990). In García Márquez's case, this need not have involved much more than the use of word-processing software and manipulation of the mouse and keyboard. For García Márquez—a perfectionist who in precomputer days would copy and recopy a page to correct a typing error—computer literacy greatly increased the speed with which he could write and produce manuscripts, and thus dramatically increased his literary output. It also allowed him to transform his method of writing from one that involved perfecting one page a day to working on longer sections in an integrated manner. Did the computer make García Márquez a *better* writer, however? Perhaps it is as difficult—and meaningless—to answer that as to answer whether the cane in the previous example made the blind man more perceptive. Let's just say that the activity of *García Márquez + computer + writing* is a different activity than *García Márquez + typewriter + writing*, involving a different writing process and increased productivity.

It is also clear that *García Márquez + computer – writing* would be an entirely different activity altogether; it is the author's skill as a writer that allows computer literacy to add so much value. In educational settings, one common problem is too much emphasis on basic computer literacy in isolation from broader skills of composition, research, or analysis. Without reference to meaningful content, goals, purposes, or tasks, computer literacy adds little value to learning. As Neil Postman (1993) put it, there are "no 'great computerers,' as there are great writers, painters, or musicians." Or, as Michael Bellino, another critic of school computing, argued, "Tools come and tools go. . . . The purpose of schools is to teach carpentry, not hammer" (quoted in Oppenheimer 1997, 62). However, just as hammering should not be taught without reference to carpentry, carpentry cannot be mastered without learning how to hold and wield a hammer properly. Similarly, comfort and fluency with hardware, software, and operating systems are not ends in themselves but are important components of broader learning goals and should be treated as such.

Information Literacy

The value of information literacy stems not just from the development of the computer and the Internet but also from a broader information society. The difficulty and importance of managing the rapidly expanding amount of information of the modern era was recognized more than a half century ago by Vannevar Bush (see chapter 1). Bush's dream to hyperlink information sources was eventually realized in a format he never imagined with the creation of the World Wide Web in the 1990s. The development of the World Wide Web, together with a host of other public and commercial online databases, enabled unprecedented personal access to information around the world—but only to those who have physical access to new technologies and the appropriate information literacies.

The skills and understandings involved in using ICT to locate, evaluate, and use information include being able to

- Develop good research questions
- Determine the most likely places to seek relevant information
- Select the most appropriate search tool
- Formulate appropriate search queries
- Rapidly evaluate the result of a search query, including the reliability, authorship, and currency of a source
- Save and archive located information
- Cite or refer to located information (See further discussion of these points in Shetzer and Warschauer 2000.)

Information literacies involve both computer-specific knowledge (e.g., mastery of browsing software and search tools) and broader critical literacy skills (e.g., analysis and evaluation of information sources).

Many of these broader critical skills were also important in the pre-Internet era, but they take on greater importance now with such vast amounts of information available online, much of it of questionable quality. Burbules and Callister (2000, 96) point out four types of troublesome online content, which they label "the 4 M's". These include *misinformation* that is false, out-of-date, or incomplete in a misleading way; *malinformation* that some will consider "bad," such as bomb-making instructions, degrading images, or other information that promotes hatred or violence; *messed-up information* that is poorly organized and presented to the point where it is not really usable; and *mostly useless information*, which is of course abundant on the Web. All these types of troublesome information exist in other media as well, but the Web presents particular perils (as well as promises) because of its lack of gatekeepers.

Consider, for example, the difference between a student research assignment before the Internet was available and today. Earlier, a high school student would gather information for a paper by checking books out of the school library. These books would have been vetted twice: once by the publisher and once by the librarian who purchased the books. With the reliability of the books' contents thus established, the work of the student was largely limited to collecting and summarizing information from a variety of library sources.

Today, a student who relies at least in part on information collected from the Internet has a much greater personal responsibility to critically evaluate sources because of the unevenness of quality and reliability of texts found there. Indeed, it is impossible to even navigate or search the Internet without making very rapid judgments as to the reliability of various sources of information. A reader must decide on the spot whether to pursue information on a particular page, follow links to other sites, or return to a search engine for another try. In such a circumstance, it makes little sense to discuss critical literacy as a separate or special construct; rather, critical literacy is an essential element of reading in the online era.[3] And, as Burbules and Callister (2000) point out, critical reading of the Web involves analyzing whether a site is credible, examining its viewpoint, asking why information is presented in a particular fashion, considering what kinds of information are left out of the presentation, and determining whose interests are served by the site's emphasis, organization, or omissions.

There is a vast difference between information and knowledge, and information literacy is crucial for being able to transform the former into the latter. Such literacy is distributed unequally in society (Hargittai 2002a) and intersects with other forms of social stratification. Promoting information literacy should be an important goal for projects seeking to promote social inclusion.

Multimedia Literacy

In the past, literacy chiefly meant text-based literacy. That is because the main technologies of literacy, such as the printing press, have "privileged the written language over all other forms of semiosis" (Kaplan 1995, 15), thus separating verbal from iconographic information and representation.

But human beings have a desire for what Jay David Bolter has called "the natural sign" (Bolter 1996, 264). As Bolter explains, "Pictures or moving pictures seem to have a natural correspondence to what they depict. They can satisfy more effectively than prose the desire to cut through to a 'natural' representation that is not a representation at all" (265–266). This desire for the natural sign—partially suppressed by the limitations of print—has expressed itself widely throughout the twentieth century in the popularity of film and television, and in recent developments in newspapers, magazines, and books. Kress (1998; Kress and van Leeuwen 1996) illustrates nicely how both newspapers and textbooks have dramatically altered in format in recent years, with visual images becoming increasingly prominent.

It is in the realm of computers, however, that multimodal communication has progressed the furthest, combining text, backgrounds, photos, graphics, audio, and video in a single presentation. The falling cost of computers and multimedia software means that millions of people around the world have the desktop power—if not necessarily the skill—to create multimedia documents, ranging from simple PowerPoint presentations to homemade movies.

This reemergence of the "natural sign" has profound implications for digital democracy. The domination of writing over other forms of semiosis has long contributed to social inequality. Learning to read and write takes years of schooling, and the gap between the schooled/literate and the unschooled/illiterate (whether at the individual, village, or societal level) has intersected with, and contributed to, almost all other socioeconomic divides of the last five hundred years. Text literacy also privileges the few dozen dominant written languages of the world (many with colonial histories, such as English, Spanish, and French) at the expense of indigenous languages, many of which lack a written form or suffer from a paucity of published material. Finally, the social practices of text literacy in schools—decontextualized, individual study and memorization rather than collective creation and interpretation—have further marginalized nonelite groups throughout the world, including many tribal and indigenous peoples whose traditional methods of learning focus on shared storytelling using audiovisual elements such as song, chanting, and dance (Warschauer 1999). For all these reasons, the rise of multimedia should provide an important opportunity to level the playing field of literacy by restoring the status of more natural forms of audiovisual communication that are in some ways more broadly accessible.

However, in other ways, the economics of the information technology industry and the social stratification of educational systems make multimedia creation highly *inaccessible* to the masses. While the cost of computers and Internet access continues to fall, the cost of the hardware, software, and bandwidth necessary to create the newest forms of multimedia remains more expensive. Stratified access to more powerful multimedia computers thus parallels other types of income and educational stratification discussed earlier in this book (see Becker 2000). In addition, in the United States and many other countries, unequal educational systems mean that students in wealthier communities get more frequent opportunities to create sophisticated multimedia whereas low-income students often are relegated to using computers for remedial drills and exercises (Becker 2000; Wenglinsky 1998). As a result, the potential of multimedia as a force for social equality can be turned into its opposite, with some sectors of the population learning how to become the producers of tomorrow's multimedia content while others are prepared only to be passive recipients (see Castells 2000b; Warschauer 1999). This discrepancy between the *potential* of multimedia literacy in promoting social inclusion and the unequal *access* to the tools and practices of multimedia literacy deserves attention.

Computer-Mediated Communication Literacy

Computer literacy, information literacy, and multimedia literacy have been widely noted and discussed by others. Computer-mediated communication (CMC) literacy has not received as much attention. CMC literacy refers to the interpretative and writing skills necessary to communicate effectively via online media. At a simple level, this includes the "netiquette" of polite online communication. At a more advanced level, it includes the pragmatics of effective argumentation and persuasion in various sorts of Internet media (e.g., e-mail, Web-based bulletin boards). At the most advanced level, CMC literacy includes knowing how to establish and manage online communications for the benefits of groups of people (e.g., community organizations running their own discussion or training sessions online).

Much basic CMC communication skill is learned implicitly and needs no instruction; an hour or two in a chat room, and a teenager will begin to pick up the style of interaction used most in that particular online space. However, it would be a mistake to infer from this that CMC literacy is developed spontaneously through social interaction or that CMC is not important since it partly revolves around chat. Indeed, in recent years, CMC has become a potent form of business (American Management Association, cited in Warschauer 2000a) and academic communication (Agre 2001b), and its more sophisticated forms are not as easy to learn. In this light, it is useful to consider the distinction made by Jim Cummins (1984) between Basic Interpersonal Communication Skills (BICS) and Cognitive Academic Language Proficiency (CALP) in relation to immigrant students to the United States and Canada. Cummins pointed out that even though immigrant children learn conversational skills in English by means of informal chatting on the playground, they still lack mastery (and need instruction in) more cognitively challenging uses of their new language, such as reading and writing academic papers, even long after they are speaking English fluently. In the same ways, the fact that children know how to chat on a computer does not mean they know how to write an effective e-mail message to a business organization, academic institution, or political representative.

The importance of CMC literacy can be illustrated through an example from academia. One important international divide in any consideration of new technologies and people's access to them concerns control of research—with academics in developed countries often usurping authorship rights of research partners in developing countries. In this particular example (Warschauer 1999), a Chinese researcher working in Shanghai had carried out much of the ground research in a study on community public health and had had prior agreement from the other research team members that a certain part of the research data would be under his control. In contrast to this agreement, though, his two Swedish colleagues e-mailed him and informed him that they were going to publish the research only under their own names. The Chinese researcher had no idea of how to write an effective e-mail message protesting this situation, and following norms of oral communication common in China, he wrote a draft of an e-mail that addressed the issue only in the most circular fashion: first devoting a lengthy introductory paragraph to discussing the health of his Swedish colleague's mother. Fortunately, that e-mail message—which, by failing to adequately reply to the Swedes' authorship proposal, could have cost him his authorship rights—was never sent. After much discussion and intervention from some American colleagues to whom he showed the draft of his e-mail message, the Chinese researcher was able to rewrite the message in a much more direct and effective manner. The Swedish colleagues were persuaded to yield and included him as author on the paper.

This last example illustrates how electronic literacy involves far more than being able to operate a computer. Rather, it is an act of agency: "the power to construct a representation of reality, a writing of history, and to impose reception of it by others" (Kramsch, A'Ness, and Lam 2000, 97). Its practice involves not only the individual activity of decoding and encoding text but also the social activity of exercising control. Like other forms of literacy, it entails not only reading the *word* but also reading the *world* and, in a sense, writing and rewriting the world (Freire and Macedo 1987).

The mere physical presence of computers will not guarantee that these important literacies are mastered. They will most often be learned in educational settings—which is the focus of the following section. First,

I briefly examine general educational theory and then, with this as a backdrop, specifically look at the role of technology in education.

The Social Life of Education

Educational debate in the United States and many other countries has been dominated by two schools of thought. The first of these views education as a transmission process. The second of these views education as a constructivist process. Both, however, downplay the social aspect of education, and it is that aspect that is particularly relevant in evaluating the potential impact of ICT on learning.

A transmission perspective considers education as the acquisition of facts, information, skills, and knowledge through a regime of lecture and tutoring (a process derogatorily referred to as "the filling of a pail"). It is the perspective behind E. D. Hirsch's widely cited book *Cultural Literacy* (1987) and his subsequent series of books listing the exact facts that students at each grade level in the United States should know. It is an approach that is clearly at odds with the imperatives of an information age, in which the memorizing of facts that our grandparents knew is much less relevant than one's ability to construct and communicate new knowledge from a wide variety of data sources (Bolter 1991). A transmission approach can be supported by technology, but only in its narrowest uses, as illustrated by an anecdote about the veteran teacher who boasted:

Don't tell me about educational technology. I've been using it for decades. In fact, I have all my lecture notes written down on a huge rolled transparency. Every year, I stand in front of the class and just roll the transparency, and the students copy down the information in their notes. And that's how I can tell how well the class is going. The faster I can roll, the better it's going.

The commonly posted alternative to the transmission perspective is a constructivist approach to education. Based on the ideas of Jean Piaget (1970), constructivism views learning as an internal mental process based on an individual's discovery of external phenomena. Constructivists seek to foster opportunities for exploratory learning and the development of mental models of how things work or are accomplished. Constructivists are strong supporters of educational technology (see Schank and Cleary

1995), especially through the use of computer programming to promote discovery learning (Papert 1980). In fact, though, these types of constructivist learning activities—such as using a programming language (e.g., Logo) to design on-screen representations or to control Lego toys—involve a great deal of social interaction, so while constructivism is obviously an improvement over transmission perspectives, it too can benefit from the types of social perspectives on learning discussed here.

Theoretical opposition to the transmission and constructivist approaches colors many of the educational debates taking place in the United States and other countries at present. A prominent example of this is the overhyped battle between advocates of *phonics* and *whole language* for reading instruction. Supporters of phonics see reading as emerging from the memorization of dozens of rules about how individual letters are sounded out. According to this perspective, these rules need to be transmitted to learners in order for learners to become fluent readers and writers. Whole language advocates, on the other hand, see reading as an emergent psychological process based on children's discovery of meaning. Similar struggles are taking place over issues related to spelling (Shall learners' mistakes always be tolerated?), writing (the following of rules or the discovery of voice), mathematics (the memorization of rules or their discovery), and science (memorization versus experimentation). Unfortunately, these debates often obscure, rather than shed light on, the actual processes by which learning takes place. In particular, the transmission and constructivist approaches both fail to fully value the social factors that are at the heart of learning and education. These social factors take place at both a microlevel of communities of practice and a macrolevel of social reproduction.

Communities of Practice

Learning is as much about enculturation as it is about transmission or discovery (e.g., Lave 1988; Ochs and Shieffelin 1984). To begin with, almost all human learning takes place within *communities of practice* (Lave and Wenger 1991). Communities of practice are networks of people who engage in similar activities and learn from each other in the process. Sometimes communities of practice are found in formal learning structures, such as classes and schools. More frequently, they are grounded in informal networks such as families or professional or occu-pational groups, and in social circles that occur in social contexts such as work or sports.

Learning in communities takes place through a process of apprentice-ship. This occurs at every level from the most basic (e.g., a child learn-ing to walk or talk) to the most advanced (e.g., a medical internship or doctoral program). Some of the learning that occurs by means of appren-ticeship is via direct instruction, for example, when a coach shows a player the right way to shoot a basketball. Much more of this learning—even in formal educational settings—occurs informally or incidentally, as learners and experts observe, imitate, experiment, model, appropriate, and provide and receive feedback. In short, an ideal learning situation provides the kind of scaffolding needed for apprenticeship learning to take place in a safe, supported way. This scaffolding might include the provision of models and resources, the organizing of learning activities in desirable sequences, and the use of conversation and discussion to tackle difficult questions.

Apprenticeship often occurs through a mentoring process with a teacher or a more capable peer. Lev Vygotsky (1978) has described this type of learning in terms of a zone of proximal development (ZPD). The ZPD is the distance between what a learner can accomplish by himself or herself compared to what he or she can do with the assistance of others; a learner advances through this ZPD by gradually taking on tasks alone that he or she previously could accomplish only with expert assis-tance. However, the involvement of an expert or mentor is not a neces-sary requirement for apprenticeship learning to take place. Informal networking among peers is also a valuable source of learning, and often more powerful than direct instruction. Learning situations that provide for a good deal of informal peer networking maximize people's oppor-tunities to learn; situations that exclude this kind of informal network-ing can endanger the learning process. This principle has a good deal of importance in relation to new technologies; consider, for example, how relatively easy it is to learn a new computer program while working in an office or other environment where others already are using it (and thus can provide ready feedback and assistance) compared to how diffi-cult it is to learn a new computer program by oneself at home.

Why are communities of practice so important? First, because the most valuable learning in society involves not so much *learning about* as *learning how* (Brown and Duguid 2000). An excellent example of this is found in writing. One learns to write not by memorizing facts about writing but by engaging in the social practice of writing in the company of colleagues, peers, critics, and mentors. Learning how to write involves appropriating the language of others, reproducing examples of writing that one reads, responding to questions and suggestions, and receiving and considering the guidance of expert critics—a classroom teacher, a dissertation committee, or the peer reviewers for an academic journal. Learning *about* writing might take place in a few days by reading a book; learning *how* to write takes many years of engagement in communities of writers.

Equally important, *learning how* is intimately tied up with *learning to be*, in other words developing the disposition, demeanor, outlook, and identity of the practitioners. This is obvious at the most advanced level of learning in universities. For example, learning how to conduct scientific research inevitably involves learning how to think, act, and interact as a research scientist. It is also true at the most basic level. For example, a study of computer users at community technology centers in California found that identity formation was a critical component of learning how to use a computer (Stanley 2001). Many of the study participants who owned home computers admitted they had never used them, partly out of fear and lack of knowledge but also because of their own self-concept. They simply didn't see themselves as the type of people who used computers. As one person noted, "I thought [computers] were too much to dream about; like a dream that is too far from reality. I couldn't see myself as someone who uses computers. I thought they were for smart people or college students" (17). Fully 70% of the interviewees in this study mentioned similar self-concept or identity issues related to computers. Only after coming to a community technology center did they begin to change their self-perception of themselves as computer nonusers. For these learners—and for so many others—effective learning involved not only a mastery of skills but also joining a community of practitioners.

Social Reproduction

Social context plays an important role in how educational institutions and processes are structured, not only at the microlevel of community interaction but also at the macrolevel. The key concept here is social reproduction; in other words, educational institutions are structured in ways that reflect and contribute to the broader social, economic, politic, and cultural relationships (Bowles and Gintis 1976; Willis 1977).

The most interesting research pertaining to social reproduction and educational technology has been conducted by Larry Cuban. Cuban's (1993) ninety-year history of educational practices demonstrated that U.S. schooling is highly resistant to reform, as teachers' behaviors are constrained in numerous ways by societal norms and expectations. Meaningful reforms that do take place in U.S. schools almost always benefit the most economically privileged students, who are deemed suitable of engaging in critical and reflecting learning. Reforms in low-socioeconomic schools generally take place on the margins of the educational process and fail to seriously transform the learning process. Cuban (1986) conducted a parallel historical study that examined uses of educational technology since 1920, including radio, television, and film. He found that technology was frequently imposed by outside parties—especially self-interested technology businesses—and had little impact on reshaping education, which responded instead to broader socioeconomic imperatives. Cuban's (2001) latest study on educational computing has found similar dynamics, with educational institutions still highly resistant to reform despite infusion of new technology into schools and classrooms. The requirements to cover curriculum, to prepare students for standardized tests, to change classes at fifty-minute intervals, and to maintain discipline and order make it difficult for teachers to engage in creative technology projects with students, except in elite schools that generally have better and more flexible teaching conditions.

Situated Learning and Critical Pedagogy

How then is a social approach to education translated into classroom practices? To accomplish this, two socially based models of teaching and learning are required: situated learning and critical pedagogy. Situated learning has two main focuses. The first is on assisting students to become part of learning communities. As Brown, Collins, and Duguid (1989, 33) explain, "To learn to use tools as practitioners use them, a student, like an apprentice, must enter that community and its culture. Thus in a significant way, learning is . . . a process of enculturation." The second emphasis is on creating relevant situations by providing students with opportunities to "carry out meaningful tasks and solve meaningful problems in an environment that reflects their own personal interests as well as the multiple purposes to which their knowledge will be put in the future" (Collins, Brown, and Newman 1989, 487). These focuses are interrelated. For example, a high school science teacher should facilitate students' entry into the community and culture of scientists by providing students with similar kinds of tasks—formulating real questions, gathering and analyzing data, developing interpretations—that they might later engage in as researchers.

Critical pedagogy shares much with conceptions of situated learning but also emphasizes the role of learners themselves in defining their own problems based on social needs and issues facing their families, communities, and others, and on confronting these problems through collective inquiry, critique, and action as part of the educational process (Freire 1994). Learners can thus confront—or at the very least, make explicit—the problem of social reproduction by analyzing, critiquing, and challenging unequal power structures as part of their learning process in school. Cummins and Sayers (1990; 1995), for example, adopted this approach to Internet-enhanced learning by promoting long-distance partnerships to identify and address important social issues as defined by learners; an example of this kind of project is discussed in the section on Project Fresa later in this chapter.

ICT in Education

The concepts of situated learning and critical pedagogy are invaluable for understanding the relationship of ICT to education, particularly when considering programs that seek to promote social inclusion for marginalized groups. Technology assists learners the most when it is not the sole or even the main focus of teaching and learning. An overemphasis on the computer per se leads to the most basic kinds of computer literacy instruction where the students may learn little more than how to make, save, and access document files. But by using the computer and the Internet to help learners enter new communities and cultures, tackle meaningful problems, and address situations of social inequity, educators can help students master the broad range of literacies required for the information age. This principle is seen in three different types of programs: computer education, computer-enhanced education, and distance education.

Computer Education

A primary means of promoting ICT access is through computer education. Community technology centers in both developed and developing countries have set up educational programs to empower socially marginalized people to learn how to use computers. Even though the computer itself is ostensibly the topic of instruction, these programs are most effective when they link with broader purposes and functions. Examples of this are seen in the work of the international Committee for Democratization of Information, based in Brazil, and the Playing2Win Community Technology Center in New York.

Committee for Democratization of Information   The Committee for Democratization of Information (Comitê para Democratização Informática—CDI)[4] is one of the largest and most successful grassroots organizations in the world directly promoting social inclusion with technology. Founded in 1995, CDI is based on the principle that the ability to use computers is central to full economic, political, and social participation in today's world. To further public knowledge of computing, it has established more than 325 community-based educational centers in nineteen states in Brazil, with a few additional centers in Uruguay, Chile, Colombia, Mexico, and Japan.

These educational centers have been established through extensive public-private partnerships. In most cases, preexisting community organizations provide the facilities and management of the center, and donations for the hardware and software are sought from the private sector (e.g., information technology firms, chambers of commerce). Teachers who work in the centers are usually chosen from the community itself, with the main criteria being social commitment rather than computer expertise (since the latter can be achieved through training).

There are many similar projects throughout the world, albeit on a smaller scale. What is special about CDI and accounts for a good deal of its success is its innovative approach to computer education. In line with situated learning and critical pedagogy, CDI does not see learning computers as an end in itself. Rather, it has worked to integrate computer education into broader social issues of concern to the communities that it serves. The general umbrella theme for this work is that of citizens' rights.

The centers established by CDI are called Schools for Information Technology and Citizens' Rights (Escolas de Informática e Cidadania— EIC). These schools have two target audiences: the poor and the socially marginalized, such as prisoners, people with AIDS, street children, indigenous communities, and landless movement members. The broad theme of citizens' rights is adapted by the local teachers and learners according to the special concerns of the particular community in which the school is located. Computer skills become the backdrop to this process rather than the main focus. Members of a landless movement, for example, may learn Microsoft Word while developing flyers, brochures, and other materials to serve their own mobilizing campaigns. A neighborhood group may learn how to use PowerPoint software by developing presentations addressing themes of drugs, sexuality, crime, and other social problems confronting the community. An indigenous community might learn how to use a database program through archiving and organizing a list of vital community resources.

I visited several CDI schools in Brazil in August 2000. At the Monte Azul Community Association complex in one of the poorest *favelas* of São Paulo, an EIC had been established to assist the economic development of the community. All the participants in the EIC were teenagers or young adults who were also engaged in other community workshops related to work force development (e.g., woodworking, electronics, paper production, weaving, doll making). In the EIC the participants carried out projects related to this other vocational training, such as developing designs for the paper products they were printing. A second

EIC that I visited was at the "FEBEM" Youth Prison, and the curriculum here dealt more directly with social issues (rather than work skills), especially those of particular importance to the incarcerated youth, such as the roots of crime, the social challenges that confronted the youth, and their chances for rehabilitation upon release. They engaged in a wide range of discussion and computer-based writing activities on these themes, and produced a twelve-page tabloid newspaper of their writings and designs. This newspaper included articles such as "Misery and Poverty," "Say No to Drugs," "Life Is not Easy," and "Liberty Is not Given by the Oppressor, but Conquered by the Oppressed." Throughout these activities, the students learned to use programs such as Microsoft Windows, Microsoft Word, and Microsoft Paint, even though the software programs themselves were not the main target of instruction. And during interviews I conducted with EIC students, the participants spoke highly of both the computer skills they had gained and the broader social perspectives they had developed.

Playing2Win   Playing2Win (P2W),[5] founded in Harlem in 1983, is the oldest community technology center (CTC) in the United States P2W not only has a major presence in Harlem but has also provided two decades of leadership for many national initiatives, including the Community Technology Centers Network (CTCNet, a coordinating body of more than 500 independent CTCs in the United States).

While P2W offers a few traditional technology courses, such as Computers and the Internet—Beginners or Office Skills: Microsoft Suite, about half its adult classes and all its youth classes are based on themes and projects rather than on specific computer applications. The goal for the center, according to P2W director Rahsaan Harris, is to allow the people of Harlem to develop sophisticated multimedia skills while engaging in projects that are of social and economic significance to the participants and the community.

For example, one course offered as part of a summer youth program is called Portraits of Harlem. Participants plan and produce professional postcards and posters based on modern or historical images of Harlem. Technical skills involved in this project include taking pictures with a digital camera, transforming old photographs to digital form using a scanner, and touching up images and creating final products with a graphics editor. Business skills are also covered in this course, including how to market, promote, and advertise a product; how to find customers; and how to determine best pricing levels in order to achieve maximum profit. Artistic skills are, of course, also honed during the life of the course, and include designing the postcards or posters as well as identifying what aspects of Harlem life are of artistic and social concern to the broader community. Ties across generations are encouraged through inclusion of old family photographs belonging to participants or their families that are judged by the group to have artistic or social value.

In another course conducted regularly by P2W, entitled Web Design Studio, participants work in teams to design Web sites for local community businesses. Technical skills such as using advanced photoediting or Web page markup software are combined with broader processes such as interviewing a client, communicating with an audience, and evaluating a final product. The project serves the participants, who develop a range of marketable skills and abilities, as well as the broader Harlem community through the development of e-commerce sites for local small businesses that would otherwise be unable to pay for the development of their own Web sites.

A third popular P2W course is called Digital College Portfolio. This is geared toward lending a helping hand to high school juniors and seniors seeking to enter university or the job market. Participants learn to create an electronic portfolio with copies of their original drawings, writings, photographs, music, video clips, and cartoons. At the same time, they learn to create and manage a Web site while also familiarizing themselves with the college application and essay-writing process.

Other courses include Eco-Science, in which students use digital cameras, photoediting software, and office suite applications to document what they see during field trips to local parks; She-Thang, in which young girls create a Web site with articles on young women of color, memoirs, poetry, editorials, and media reviews (published on the HarlemLive Web site; see chapter 4); and Percussion Digital Music, in which participants learn how to digitize music by means of recording, processing, and mastering while acquiring basic proficiency in audio processing and arrangement/layout.

The pedagogy in all these courses is deliberately student-centered and project-based. This does not mean that participants are left to their own devices; rather, the instructors, guest lecturers, and, at times, more advanced students provide guidance in and assistance with technical skills and processes at the point of need.

Computer-Enhanced Education
Groups like CDI and Playing2Win were established in order to provide computer skills to low-income and marginalized communities. These organizations then developed, and continually adapt, a curriculum that they felt could best combine computer skills with other content that served community needs.

Computer-enhanced education starts from a very different premise. In computer-enhanced education, a broader curriculum is already established, based on courses and content in areas such as mathematics, science, social studies, and language arts. The challenge there is not to create an entirely new curriculum but to make effective use of technology to enhance broader educational purposes.

The main site of computer-enhanced education is public schools and colleges throughout the world. These educational systems, with hundreds of millions of students, represent a critical arena for combating marginalization from the information society. If public schools help compensate for unequal access to computers at home, they can provide an important means for promoting social inclusion and equality. If, on the other hand, schools offer unequal access to and use of technology, this can serve to heighten social stratification.

Computers in U.S. Schools   In developed countries, educational use of computers has the potential to either help overcome or worsen social stratification. On the one hand, technology can be an equalizing force, by giving all students access to a tool/medium that is vital for today's education. On the other hand, if technological resources are unequally distributed or used in schools, ICT can serve to stratify already existing inequalities.

The United States is the country with perhaps the longest and most extensive use of computers in education; nevertheless, the results to date are not encouraging. Evidence suggests that the use of computers in education is tending to worsen rather than help overcome societal inequality.

There is differential physical access to computers and the Internet in schools in relation to income and race. According to data gathered by Market Data Retrieval, schools in high-poverty communities in the United States have one computer for every 5.3 students compared with a ratio of 4.9 in low-poverty communities ("Dividing Lines" 2001). The difference is even greater at the level of Internet access: high-poverty schools have a ratio of 9 students to 1 computer with Internet access compared with a ration of 6 to 1 in low-poverty schools. At the economic extremes, the differential is still greater (Cattagni and Westat 2001). Or, to put it another way, according to the Department of Education, in schools that have fewer than 11% of students living in poverty, 74% of classrooms have Internet access ("Dividing Lines" 2001). On the other hand, in schools with 71% percent or more of students who are poor, only 39% of classrooms are connected to the Internet.

Differential access *across* schools is multiplied by differential access *within* schools. In a study of 185 primary, middle school and high school teachers in an urban school district, Janet Schofield and Ann Davidson (2000) found that some 25% of the teachers engaged in Internet activities primarily or exclusively with academically advanced students, but only 5% carried out Internet activities primarily or exclusively with lower-achieving students. And the differential access continues at the *within-class* level. A full 70% of the teachers in the Schofield and Davidson study only allowed some (rather than all) of their students to use the Internet, with the privilege usually reserved for those who were already performing well academically or behaving especially well. Internet use in the schools involved in the study appeared to be regarded by teachers as a reward or special privilege for model students.

Of course, much more important than the actual amount of time that students spend on computers and the Internet is what they actually do there. Here the evidence is especially discouraging. Basically, a disproportionate number of poor and Black or Hispanic students are engaged

Table 5.1
Primary Computer Uses by U.S. Eighth-Grade Students (as Reported by Their
Teachers)

	Simulations/Applications (%)	Drills/Practices (%)
U.S.	27	34
Asian	43	27
White	31	30
Hispanic	25	34
Black	14	52
High income[a]	33	31
Low income[a]	22	34
Private school	30	10
Public school	27	36

Source: Adapted from Wenglinksy (1998).
[a] High income and low income determined by eligibility for subsidized school
lunch.

by their teachers in using computers for remedial drills, while well-to-do
and White or Asian students significantly more often use computers for
applications and simulations promoting higher order thinking (table
5.1).

Paradoxically, it seems that because of the popularity of remedial drill-
and-practice (more pejoratively known as drill-and-kill) software, low
socioeconomic status (SES) students often end up using computers in
classrooms more than high SES students, at least in middle and high
school settings. For example, a national survey by Henry Becker (2000)
showed that in English and mathematics—the two subject areas in which
remedial software appears to be most popular—low SES students are
reported as using computers more frequently than high SES students. In
science, a subject in which many computer-based simulations and other
advanced applications are available, computer use is reported as being
greatest among high SES students.

Recently, over the course of an academic year, I conducted observations and interviews on a regular basis in two high schools in Hawai'i:
one was an elite, expensive private school in a well-to-do neighborhood,
and one was a rural public school in an impoverished community

(Warschauer 2000b). Students in the former were principally white or Asian-American from middle-class or wealthy families. Students in the latter were overwhelmingly native Hawaiian or Pacific Islanders (e.g., Filipino, Samoan) from low-income families. Both of the schools had a reputation for making excellent use of educational technology and were in fact selected for the study for that very reason.

Interestingly, both of the schools had engaged in the kinds of reform that are portrayed as critical to effective use of technology (see Means 1994; Sandholtz, Ringstaff, and Dwyer 1997). These reforms served to devolve more power and authority to students and to move away from the kind of lock-step teacher-centered instruction that is often criticized. For example, both schools used a good deal of student-centered project-based work, especially in the technology-intensive courses. Both schools allowed for flexible scheduling, so that students could sometimes meet for two or more hours in a block of time rather than only in fifty-minute periods. And both schools made use of interdisciplinary team teaching, so that subject teachers across disciplines could work together to help coordinate interesting projects.

In spite of the extensive use of technology at the two schools—and the common reforms that in theory should have allowed technology use to be effective—the ends to which computers were put were markedly different when the two schools were compared. Simply, regardless of the subject area, the elite school used technology to help prepare scholars, whereas the poorer school used technology to help prepare people for the work force.

To provide an example, students in a biology class at the elite school used hand-held devices to probe the temperature, acidity, and absorption spectra of plant life in nearby ponds, downloaded the data to personal computers, and then used a software program to graph, compare, and interpret the data. This project was team-taught by the biology and the math teacher so that the students could learn to make better use of calculus algorithms in their data analysis. In contrast, students in a marine biology class at the poorer school used computers to edit a newsletter about their personal experiences with marine biology, discussing, for example, their personal feelings about a class trip around the island of O'ahu. The students also wrote and compiled inspirational

DOC

stories from their own lives and produced their own local version of the well-known book *Chicken Soup for the Soul* (Canfield and Hansen 1993). These activities were unquestionably educational, and certainly involved computers and a range of applications, but they did not focus on the discipline of science per se, as was the case in the wealthy school. Perhaps even more telling, instead of the team-teaching partnering involving biology and math, in the low SES school the biology and business teachers collaborated on projects. The former partnership was designed to strengthen students' skill in analysis of complex biological phenomena whereas the latter pairing emphasized the vocational aspects of the course.

The teachers involved in these two classes were quite explicit about the different goals of their instruction. The biology teacher at the elite school—a former research scientist at Stanford—explained how he deliberately geared all his in-class activities, including computer work, toward helping students develop the research skills necessary to enter college. As he explained,

We've been working over the years on our biology program, particularly our advanced biology program, to give students the type of experience that they need to prepare them for college work. So I have a very strong background in research, which I loved. And I try to share that love of research with my students. And since I was pretty much lab oriented and biochemistry oriented, I did what I knew and tried to implement those kinds of experiments. And it became obvious, over the last ten years, that computers were becoming one of the most important scientific tools available. And, so we wanted to implement the computers into the program. We realized that this was an important scientific direction for our students to go.

In contrast, the science teacher at the poorer school, who was not herself a science major, saw her goal as helping prepare her students for employment:

I looked at these kids and I said, how many of them are going into a science-related field? Out of fifty, it's lucky if three or four of them would go into something science-related. So I said, this is really not acceptable. And that's where I changed my focus. As far as I'm concerned they don't have to learn the science or learn the material, as long as they're doing these projects. But my focus is on them being respectful, responsible, and seekers of information. And I said, then I don't care what you do, whether you go out and be a trash collector or dig ditches or if you go into a community college.

The teacher justified her approach by citing information that had been gathered from local employers, who had been surveyed by the school and who had said in a response to a survey question asking what background knowledge they expected, or most wanted, their employees to have, "We don't care what they [the students] studied, we want a student who's respectful, who's responsible, who can work together with other people and want to learn, we can train them. We don't care. We don't need them to be honors students and all that. We can train them on the job. Give us kids who know how to be respectful, responsible, team players." It seems that the teacher's approach in class, including her use of computers, was a direct response to potential employers' expectations of her students rather than a concern with teaching science and producing students who knew how to conduct science like scientists do in the world outside school.

I have treated this example in depth not because it is special but rather because it is common and illustrative; it reflects not only what other teachers at the two schools are doing but what has been observed more commonly in education elsewhere (e.g., Becker 2000; Cuban 1986; Warschauer 1999). Technology by itself does not change the nature of schooling. At its best, it amplifies processes that are already under way. A school system that channels students into different social futures through a variety of mechanisms—including unequal funding, disparate teacher preparation, and a tracking system—will not be turned around by computers. It is more likely that the channeling effects of such a system will only be amplified.

Perhaps the most interesting part of this particular example is that the use of technology in the low SES neighborhood school was found not in drill-and-practice software but in collaborative writing and communication tasks. Indeed, the production of limited-content newsletters or Web sites was common in other technology-enhanced courses in the school and appears to be a rising trend nationally. I would suggest that this is in line with broader economic changes in U.S. society. In an era when basic computer literacy and teamwork is a requirement of almost all office jobs, more classrooms in low SES schools will likely focus on these skills. We thus might see a day in the future when across the board more low SES students are using the Internet than high SES students, but for the kinds of tasks that prepare people for basic office work rather than for scholarly pursuits.

Of course, an emphasis on vocational preparation in this school or other schools is not in itself wrong. Indeed, better preparing students for the work force is a widely accepted goal, and it may be especially important in communities with high levels of unemployment and poverty. And elsewhere in this book I highlight other educational efforts that are designed precisely to provide teenagers or adults with better job-related skills. However, this was not a vocational school but a comprehensive high school with the supposed mission of preparing students not only for the workplace but also for university studies. At this school, though, as in many low-income neighborhood schools in the United States, academic opportunities were constrained not only by low expectations but also by very difficult working conditions. Even the teachers at this school who wanted to offer more academically challenging projects had difficulty doing so because of large class sizes, lack of modern equipment, and insufficient time for preparation. Thus the effect of technology infusion, even in a school with talented, well-intentioned educators, was shaped and constrained by the broader social context.

However, while social context shapes what takes place in classrooms, it does not completely determine it. Excellent examples abound of teachers in low-income neighborhoods finding ways to use technology to support academic achievement. Two examples are discussed in the next sections.

Project Fresa: Fostering Critical Thinking with Technology   Mar Vista Elementary School is located in the midst of strawberry plantations in Oxnard, California, a couple of hours' drive north of Los Angeles. About 80% of the students in the school are Latino (including Mexicans, Mexican Americans, and Latin Americans), and the majority of them have family members working as laborers in the strawberry fields. Even though most schools in California have ended formal bilingual education following a 1998 statewide initiative, Mar Vista is one of a small number of schools that have continued their bilingual programs thanks to a progressive administration and parental demand. Teachers at Mar Vista have also become leaders in effective use of ICT to promote academic skills and critical awareness among traditionally marginalized students. This is accomplished through theme-based project-oriented instruction that is sensitive to students' own social concerns while at the same time engaging students in complex and cognitively demanding learning tasks.

An example of this is Project Fresa,[6] a theme-based year-long project for primary school students. The project takes as its main focus the local strawberry (*fresa*, in Spanish) industry. The children begin by formulating their own research questions about the conditions of strawberry workers. They then use these research questions to generate interview and survey questions, enrolling their family members, relatives, and neighbors as respondents. (They often conduct the interviews in Spanish and then translate responses into English.) Afterwards, the students learn to record in spreadsheets and to produce graphs in various formats of the data they have gathered (analyzing, for example, which types of graphs best display which types of information). The graphs are incorporated into PowerPoint presentations together with photos and quotations from the people they have interviewed. With the guidance of the teachers, they then search on the Internet for further information about the conditions of strawberry workers and also invite guest speakers into their classroom from environmental and workers' rights groups. Based on the information obtained from the Internet and guest speakers, students write letters via e-mail to the strawberry growers, expressing any concerns they might have about strawberry workers' rights. In past years they have also sent e-mails to elected officials, such as the state governor, with real and informed inquiries about agricultural laborers' rights. After engaging in this kind of work, they begin an e-mail exchange with children in Puerto Rico who live in a coffee-growing area, to compare notes about the two industries and the condition of workers. At the end of the year, the students at Mar Vista hold a public presentation, to which their parents and other community members are invited, to display the multimedia products they have created.

Compared with using the computer for drills and exercises, this kind of project-based teaching has several strengths. Students learn to actively master technology rather than use it in a passive manner. They engage in their own research, data collection, analysis, and interpretation, and produce high-quality products such as letters to elected officials and data-based presentations. They also learn to speak out and take action on issues of importance to their community. Through gathering and evaluating information from a variety of sources, including workers, nongovernmental organizations, businesses, and politicians, students involved in Project Fresa gain a better understanding of how different players shape the strawberry industry and the conditions of its workers.

Technology Academy at Foshay Learning Center Foshay Learning Center[7] is an urban K-12 school in a socially and economically distressed neighborhood of central Los Angeles. Some 72% of the students in this school are Hispanics and the remaining 28% are African Americans. Almost all the students are from low-income families. The surrounding area is known for a high degree of unemployment, crime, and drug use. The school, under the leadership of the (recently retired) principal Howard Lapin, has succeeded in creating and maintaining well-organized programs that emphasize academic achievement. The school has won numerous national awards and was designated one of the best 100 high schools in the nation in the year 2000.

The high school portion of Foshay is divided into three academies: Finance, Health, and Technology. All students in the high school choose one of the three academies in which to locate themselves, based on their career interests and aspirations. While computers and the Internet are well integrated into all the programs of all three academies, students in the technology academy develop special expertise in this area. Within the technology academy, courses in the first year (grade 9) focus on topics such as programming and multimedia authoring. Students spend a good deal of time working on their own projects that focus on a variety of social themes. Some of these projects also involve developing authentic materials for the community, for example, developing promotional brochures for local nonprofit organizations. Others involve research projects on local social conditions.

The technology-oriented topics focus on genre, message, and purpose as much as they do on technique. For example, in learning to use PowerPoint, students discuss and practice specific presentation genres, such as a stand-alone PowerPoint presentation (where presentation is shown without a speaker and thus must be complete in itself) and a support presentation (in which the PowerPoint show serves to supplement what is said by a speaker). Through a number of individual and collaborative assignments, students develop a sense of how to use presentation software to communicate effectively. These lessons are then hammered home through interdisciplinary projects involving teachers from several content areas. For example, a social studies, English, and technology teacher might combine to organize a research and presentation project on a topic such as the conditions of immigrants in Los Angeles. The final Power-Point presentation would be evaluated according to its content, language use, and technique through collaboration of the three instructors. By the end of their last term of high school, graduates produce a CD-ROM with a portfolio of all their multimedia work completed over the four years of their high schooling for use in their college applications or job searches. Internships at local technology firms during this final term also provide students with more direct opportunities to learn about the role of technology in society as well as to engage in meaningful commercial projects.

In summary, use of technology in U.S. schools appears to have a mixed impact on social inequality. In many circumstances, technologization of schools has worsened educational and social divides. However, outstanding examples exist for ways to integrate ICT to promote social inclusion. These involve not only a reorganization of the classroom but also a pedagogical approach based on critical collaborative inquiry and analysis.

Educational Technology in Developing Countries   Developing countries must weigh two divides when they consider integrating computers into schools. On the one hand, failure to technologize their schools and their societies can lead to a heightened *international* divide, as they fall more and more behind technologically advanced countries of the developed world. On the other hand, too great an emphasis on technology, at the expense of more basic educational problems such as building and resourcing primary schools in rural areas, can lead to a heightened *national* divide between rich and poor or urban and rural.

The solution to this contradiction is to introduce ICT in measured ways into schools, through carefully designed pilot programs. That way, the country's educational leaders can best learn lessons about what works in particular national and local contexts, and better plan for larger-scale spending on technology as computer and Internet access prices continue to fall.

This approach, however, is often easier said than done, as illustrated through the following examples of the introduction of educational technology in two developing countries.

Egypt   The government of Egypt has devoted substantial resources to the area of educational technology in recent years.[8] Through computerization of public schools, the government hopes to better prepare the Egyptian people for a technology-based global economy, and thus to help the country reach the educational, economic, and social levels of more developed countries.

The government has backed its commitment to educational technology with substantial resources. A national plan for the technological development of education was initiated in 1994 and soon thereafter a national Technology Development Center (TDC) was formed in the Ministry of Education. Since then the TDC has grown to over 600 full-time staff who have the responsibility for coordinating and implementing technology projects throughout Egypt's governmental school system.

The TDC has placed multimedia rooms in all secondary and middle schools, and many primary schools. These rooms comprise one to two high-end computers, a video projector, an array of software, and an Internet connection. In addition, all secondary schools and many middle schools have computer laboratories, with ten to fifteen computers running on MS-DOS or Windows platforms. At the same time, much of the Egyptian school curriculum has been transferred to CD-ROM format, and computer training programs have been established by the government for Egyptian teachers.

Unfortunately, the results of this substantial investment have been disappointing. The computers in the multimedia rooms—which effectively means only one or two computers per school—are spread too thin to make any real difference to learning. In any case, the rooms are often kept locked because local school authorities don't want to suffer the risk of having expensive equipment damaged through use by teachers. This phenomenon has been reported frequently in the Egyptian press. As one article ("PCs and Teachers" 2000, 2) exclaimed, "Primary school teacher Hasnaa el-Hefnawi is enraged by the decision to introduce the computer science curriculum. . . . The ministry has repeatedly tooted its own horn about how many computers it has supplied to schools. 'Doesn't the minister realize that these computers are kept in school warehouses like antiques or used merely for decoration?' she mused."

On the occasions when students do make use of these multimedia rooms, they generally watch the teacher lecturing, as is the normal practice in Egyptian education (Tawila et al. 2000), but this time with the aid of a CD-ROM and projector. The CDs themselves contain the same material as the students' textbooks, transferred to a new medium. With very large classes, poorly trained teachers, and a curriculum geared toward test preparation, there is little in the Egyptian educational system that would support more creative use of the multimedia rooms.[9]

Meanwhile, the computer laboratories are used almost exclusively for a course in basic computer literacy, which focuses on mastering DOS and Windows commands. Teachers of that class, as of other classes, are not allowed to depart from the prepared curriculum, nor are they prepared to, based on their knowledge, background, or training. The laboratories themselves, which could potentially offer a site for creative hands-on use by students in other subjects or after school, are not allowed to be used for anything other than the specified computer literacy courses.

Ministry of Education training courses for teachers similarly focus on basic computer literacy, such as how to turn on a computer, how to operate basic programs, and how to use a CD-ROM. No in-servicing is available for teachers concerning new pedagogical approaches involving technology.

Finally, Internet access in schools is routed by telephone via the Ministry of Education Offices to ensure better control. This necessitates a double-connection process that rarely functions. In any case, only the staff person in charge of the multimedia room in each school is given the Internet account information; neither teachers nor students are allowed to access the Internet on their own.

Meanwhile, the large sums spent on educational technology drain resources from Egypt's urgent need to improve its primary education. With a literacy rate of only 54.6%, and only 42.8% for women (UNDP 2001), Egypt's poor primary education is a major drag on the country's economy (Birdsall and Lesley 1999; Fergany 1998). Egypt already suffers from one of the more unequal educational systems in the world, with proportionally far too many resources going toward university and secondary education, which benefit only a minority of the population, as opposed to primary schools (Birdsall and Lesley 1999). Expenditures on technology, which are disproportionately devoted to secondary and tertiary institutions rather than to primary schools, are worsening this gap while thus far bringing no appreciable benefits.

Using technology to transform or even improve education is exceedingly difficult, and much wealthier countries such as the United States, have also had substantial difficulties. According to one international comparative study, almost every country goes through the same learning experience in implementing educational technology: first focusing on computer drills, then computer literacy, and only later learning the value of emphasizing real applications (Becker 1993). In other words, there are steep learning curves in the area of instructional technology for government officials and educators alike, and it is not surprising that results in Egypt have not matched expectations. Nevertheless, there are important lessons to be gained from the Egyptian experience that can be of value for any developing country that seeks to make use of technology to promote greater development and social inclusion.

First, computer resources that are spread too thin will have little impact. It is more effective to design narrower pilot programs that concentrate resources than to democratically sprinkle a small number of computers throughout the entire national system without any realistic plan to make effective use of them (Osin 1998).

Second, as emphasized throughout this book, it is counterproductive to place too much emphasis on physical resources without attention to the digital, human, or social resources that make effective use of technology possible. In this case, the Egyptian Ministry of Education did not develop original computer-based educational content or software for either teachers or students; did not engage in the kind of training programs that would have assisted teachers in making effective use of technology; and did not create communities or social structures that could have supported good use of technology. For example, simple technologies such as e-mail were not used to provide networked discussion and exchanges among teachers or administrators. The testing system that rewards students for memorization of rote material rather than creative or critical thinking was left in place. And steep vertical hierarchies within the Ministry of Education, which stifle opportunities for innovation or experimentation at the local level, were not challenged (and were in fact duplicated within the Technology Development Center itself). For all these reasons, it is not surprising that computers in Egyptian schools have been principally used to reinforce ineffective educational practices rather than help transform them.

There are many creative minds in Egypt working on issues related to education, technology, and social inclusion. For example, the Regional Information Technology and Software Engineering Center, a semipublic institution with national and international funding, has launched a large number of innovative 21st Century Computer Clubs in low-income neighborhoods throughout the country. Thousands of poor children have developed computer skills in these clubs, and some have gone on to participate successfully in international events such as the ThinkQuest competition for children's development of academic Web sites. The contrast between the success of this private initiative and the difficulties faced in public school settings demonstrates the importance of institutional context for constraining the access and use of technology.

China   Some developing countries have had better success in launching and sustaining effective educational technology programs in public schools. A key to success appears to be well-designed pilot programs with a strong emphasis on curriculum development, teacher training, and pedagogical support (see Osin 1998). A national pilot program in Chile, called *Enlaces* (connections) received a positive review from the World Bank Education and Technology Team (Potashnik 1996). A similar program in Costa Rica seems off to an excellent start because of its strong emphasis on community participation and buy-in, teacher training, and use of technology to support broader curricular goals (Verdisco and Navarro 2000). Another interesting initiative is taking place in China, where the national Ministry of Education launched a national pilot project to investigate the possibilities of using new technologies in education.[10] The objectives of the project are to develop and try out new modes of classroom teaching with technology, to train teachers in integration of technology and education, and to generate a database of materials for computer-enhanced learning. The pedagogical underpinnings of the project are considered to be constructivism, meaningful learning, and student-centered learning.

A national core group led by Beijing Normal University, the national teachers college connected to the Ministry of Education, is working to develop curriculum and to oversee the project. A total of 600 experimental schools distributed throughout the country are participating in the program. Each of these schools has been equipped with a large, modern computer laboratory with Internet connections. Regional training centers also have been established in each of China's provinces to provide more direct supervision of the implementation in the individual schools. Each participating school sends a core group of two or three teachers to the center to participate in a ten-day initial training course that covers basic technology applications, theories of educational reform, and instructional design. The teachers then have follow-up training sessions and meetings at the regional centers and at their school sites.

Uses of technology consistent with the overall student-centered approach grounding the project are encouraged. These include teaching students to search for information, to actively master information technology tools, and to engage in autonomous learning. Beyond that, each school is given a great deal of autonomy in developing its own approaches to using the computer laboratory. In this way, the project organizers hope to encourage experimentation to see what materials, curricula, and pedagogical practices emerge that are effective in the Chinese context.

For example, NanTou primary school in ShenZhen city in southwest China has developed computer-based reading and writing materials to try to speed up literacy development. Learning to read and write Chinese characters is painstakingly slow and, when taught via traditional methods, monotonous because it involves large amounts of repetitive character and text copying. The school is experimenting at present with computer-based reading passages written in Mandarin Chinese, with glosses and dictionaries, as well as with involving young learners in computer-based writing of their own stories composing directly on the screen. Online supplementary activities include finding and reading fairy tales from Web sites and discussing them with other children through the writing of messages on online bulletin boards. Early results suggest that children are learning to write via computer much faster than they do by hand, and their creative writing activities are helping them learn to read faster (He and Wu 2001).

The project leadership team currently is researching the impact of the pilot project and is gathering data and information about best pedagogical practices that have arisen in response to placing computers in the schools. A CD is being developed by the project leadership with presentations introducing the overall educational approach, an instructional design plan, samples of courseware, supplemental instructional resources, teacher training video material (based on video clips of outstanding teachers), and reports on the evaluation process to date. The CD will be distributed to all the schools in China as a first step to try to promulgate the lessons learned from the pilot project.

The program is still in its early stages and is not without its serious challenges. As in Egypt, a test-driven curriculum provides a disincentive to develop student-centered modes of learning, especially in secondary school when students are preparing for exit exams. Though facing challenges, the overall design of the program—with a balanced emphasis on hardware, curricular resources, teacher training, and evaluation; with an initial implementation in a select number of schools; and with local experimentation and innovation actively encouraged—makes it likely that valuable lessons will be learned.

Distance Education

A final education focus related to technology and social inclusion is distance learning. In many people's eyes, Internet-based distance education shows the promise of obliterating obstacles of time and space, bringing educational resources to many people who previously lacked them.

Unfortunately, it does not always work out this way in practice, as seen by several studies of learning at a distance.

Network Science One of the most ambitious attempts to promote and document learning through Internet-based distance communication has been a set of projects known as network science. Launched in the United States in the 1980s, network science projects involved teams of children in classrooms throughout the United States and the world. The idea behind the program was that children would learn through collecting scientific data and sharing it on the Internet, providing a wealth of scientific information to promote constructivist learning. Typical network science projects involved measuring the acidity of local rainfall, tracking migrations of birds, or recording local weather conditions. In these projects, online information developed by the national or international project organizers provided instructions and supplemental materials. Chat rooms, bulletin boards, discussion forums, and e-mail lists provided opportunities for long-distance interaction.

Though these projects were presumed to support constructivist learning in the classroom, no one had really measured their impact until the late 1990s, when a research organization known as TERC carried out a five-year study of network science programs. TERC had been instrumental in launching and leading several network science programs, including some of the ones under investigation, so presumably the research scientists were hoping to find positive results. However, their final report (Feldman, Konold, and Coulter 2000) offered a devastating critique of the normal practices of network science. They found three main trends. First, students tended to upload data to the Internet without even bothering to download others' data. Second, when they did download data, they often had no idea how to analyze or interpret them in any meaningful way. And third, although the students reported that they enjoyed communicating with other students online, it was found that this interaction was usually about personal and social issues and had very little to do with science.

Some network science projects were successful, but only in cases where strong teacher mentoring, guidance, and instruction were taking place *inside* the classroom. The readings and instructions provided online were in themselves shown by the study to be ineffective in teaching children how to do science. Classrooms that depended principally on these online resources benefited little. But in classrooms where there was a very strong in-class component, with students learning how to collect, analyze, interpret, and discuss data before they ever went online, the Internet-based communication and resources added additional value. In other words, the central feature enabling effective use of Internet-based materials and distance communication was a strong local teacher working closely with students in face-to-face communication.

It is worth quoting from the summary of the five-year study:

> The experience of the network science curricula to date has led us to doubt that virtual communities for K to 12 students can replace classroom-based communities. Our reservations are based on how difficult it has proved to get substantive discussions going among participating classrooms. These reservations have been reinforced by our analysis of class discussions. . . . Given the timing, monitoring, nuanced voice, eye contact, and on-the-spot decision making required to engage students in reflective discussions, online discussions are a poor substitute by comparison. Most simply, the necessary subtleties of face-to-face interaction have no sufficient analogue online. It would be especially unfortunate if, in our ardent attempts to help classrooms get online discussions going, we inadvertently undermined efforts to improve the quality of class discussions. (97)

Finally, the research by Feldman and colleagues made it clear that distance communication faced the same limitations when used with teachers as when used with students:

> When faced with the costs associated with inservice teacher development, the possibility of using the Internet becomes an attractive alternative. However, the same critique offered earlier for why the Internet is ill suited for students learning the subtleties of substantive discourse applies as well to the challenge of teachers learning new pedagogies. The Internet may come to play an important role in sustaining contacts and building on experiences that take place in [face-to-face] summer institutes, but it is a poor substitute for such experiences. (98)

Advanced Placement Instruction   Additional education programs, some of them designed explicitly to overcome digital and social divides, have yielded similarly disappointing results. For example, in the United States, most high schools in poor communities offer fewer Advanced Placement (AP) classes than do schools in rich communities. These classes are important because they provide high school students with college-level instruction and credits that they can use to offset time needed to be spent studying at college, and they also allow students to potentially raise their high school grade point averages (since many universities, in evaluating applicants' transcripts, award a maximum of 5 grade points for AP courses rather than the 4 points given for standard high school courses). Students who have taken and passed AP classes will thus have an advantage in applying for admission to elite universities compared with students who have not. At the University of California, Los Angeles, for example, the mean grade point average for entering students in 1999–2000 was 4.15 on a 4.0 scale, meaning that even applicants with perfect grade score averages would fall behind the mean without access to extra-value AP classes. Yet access to these courses is dramatically unequal; for example, Beverly Hills (California) High School, with 9% blacks and Hispanics in one of the wealthiest cities in the United States, offers thirty-two AP classes. In contrast, at similarly sized Inglewood High School, with 97% Black and Latino students in a low-income community merely twelve miles away from Beverly Hills, only three AP classes are offered (Wales 2001a).

In order to overcome the disadvantage to students of having access to fewer AP classes, a pilot program was established to offer AP instruction online in a low-income community in California (Wales 2001b). However, the pilot program was a failure, with 73% of the students who enrolled in the AP course dropping out before the course finished. Lack of personal, face-to-face contact was a principal reason given by students for dropping out. As one student explained, "The [online] teacher in this class is not like the regular classroom teacher who can be there with you every day when you need an answer or input." Another added, "Even though there was a [online] teacher there to help us, I felt that I might have stayed in the class if it were taught more by the teacher because I learn better from a [face-to-face] teacher rather than reading everything" (Wales 2001b).

A second iteration of the program was offered students in a continuation of the pilot program, but with much greater social support. Rather than taking the online AP course in isolation at home, students instead came to school, where they took the course in the computer laboratory with a teacher from the school present in the lab. Though the course was taught online by an expert AP instructor in another city, a local teacher was on hand in the computer laboratory to help students with computer problems and other matters. In addition, students received social support from their peers. Though this setting mitigated some of the advantages of anytime/anywhere learning, it was dramatically more effective for this group of low-income youth, with the dropout rate falling from the previous 73% to 14% (Wales 2001c). In this case, and perhaps in many more, a socially supportive environment easily trumped the advantages of 24/7 learning.

Teacher Training in Brazil   Distance education programs overseas have shown similar patterns. For example, in Brazil, an Internet-based teacher training program was established to reach educators outside the major cities. The first time the program was run, 46% of the participants had dropped out by the end of the program and, for the remaining participants, much of the discussion online was dedicated to technical problems (Collins and Braga 2001). In later sessions, occasional face-to-face meetings were mixed in with the Internet-based program, and a 24-hour in-person online help service was added. As a result, the dropout rate decreased to 8% and the content of the interaction focused much more on pedagogical issues.

The Challenges of Distance Learning   What lessons can be learned? Why do distance education efforts show such difficulties, and what are the implications for issues of social inclusion? The cases presented in this chapter so far suggest that online-based education is still beset by a range of delivery problems that impinge on learning opportunities.

Superiority of Face-to-Face Communication   The first factor is the relative inferiority of online communication compared to face-to-face communication. Online communication has many advantages, as discussed in chapter 1, such as allowing fast-paced written interaction between people around the world. These advantages come into play best in particular circumstances, as when scientists discuss ideas on a discussion forum or a person with a disease seeks online information from others with similar problems.

However, in many other circumstances, face-to-face communication is superior to online communication. It is even more fast-paced and flexible than online communication, and it allows for the quick interpretation of gestures, facial expressions, and other audiovisual clues from dozens of people simultaneously as well as the quick and easy reference to drawings, charts, or physical artifacts. This rich human and physical environment allows students to better follow what a teacher is saying, and allows a teacher to quickly diagnose how students are following a presentation and thus make rapid adjustments to get a point across better. A warm smile or even a pat on the back from the teacher provides important affective support. The teacher can also rapidly break a class into pairs, small groups, or large groups, and easily have students move around the room to show each other their work and discuss ideas. While all these types of interaction can be simulated in various ways via computer, those computer-based interactions generally take much more time and are not nearly as rich in communicative content.

Value of Informal Networks   Consider for a moment why a professional conference can be so valuable. Conference attendees may occasionally benefit from what they hear in formal papers and presentations, but they much more often benefit from informal interactions outside scheduled presentation sessions. The personal chat after a presentation, the morning coffee with a new or old colleague, and the chance encounter in the hallway or elevator all provide invaluable resources for the expansion of ideas and contacts.

Students receive similar or even greater value from their social environment. A tremendous amount is learned through informal interaction and social contact (Brown and Duguid 2000). First, casual interchanges can often alert students to answers to practical problems, such as how to operate a particular computer program. Second, informal encounters help introduce learners to new ideas that they may not have been aware of. Third, and perhaps most important, learners can better understand the outlook of their instructors by engaging in talk outside the formal classroom context. Think, for example, what a graduate student gets from close personal interaction with faculty members. By dropping in at professors' offices and chatting with them at will, the graduate student not only engages in conversation but also sees what papers are on the professor's desk, what books and journals are on the professor's shelves, and learns who else the professor might be talking to over the phone or in person. In this way, the graduate student comes to learn what it is like to be a professor and can better envision himself or herself becoming one, or at least carrying out similar academic work.

Two studies from the workplace—both discussed in Brown and Duguid's remarkable book *The Social Life of Information* (2000)—show the value of informal interaction for learning. One study examined how photocopy repair people carried out their work. The study revealed that the formal printed documentation they received was almost useless in helping them solve real problems on the job. An anthropologist who studied the repair workers throughout the day (Orr 1966, cited in Brown and Duguid 2000) discovered that it was over breakfast that they really engaged in learning—when they discussed over coffee the types of work-related problems they encountered and how they actually handled them. In a second situation (see Whalen and Vinkhuyzen 2000, cited in Brown and Duguid 2000), telephone support employees learned little in their formal training sessions. However, when they sat near other more experienced colleagues in their daily work, they observed firsthand how the veterans took calls, asked questions, and gave advice. If they didn't understand the answers given, they immediately walked over and had the person demonstrate the answer, using both the desk computer and the training manual. These types of highly valuable informal interaction, whether for workplace training or academic advancement, are not likely to be provided via online learning.

Economics of Online Instruction   Finally, the economics of online instruction pose serious challenges to high-quality education. There are limits to how many students can fit in a classroom, but there is almost no limit to how many students can take an online course. Indeed, one reason that many university administrators are seeking to expand online education is for cost-saving purposes. By increasing the number of students per instructor and reusing online materials again and again, universities hope to make enormous savings. And universities are under a great deal of pressure to make these changes under competition from rapidly growing for-profit institutions that use online instruction and other types of technology to reduce their costs (Noble 1998a). Yet, it is precisely the more expensive types of online instruction—involving a high degree of professor-student interaction—that are proving the most effective (Feenberg 1999a).

All of these characteristics of online instruction have special significance for groups that are socially or economically marginalized. For example, in the United States, black, Latino, and low-income students already face severe challenges to succeed in the educational system. They have the weakest social networks outside of school to support advanced academic achievement because proportionately fewer of their relatives or neighbors have themselves received university education. They thus need the greatest support possible in schools and universities to succeed. This support should include rich face-to-face mentoring from teachers, counselors, and administrators, and plenty of informal interaction in dense social networks of peers and professors. Yet, because of economies of scale, it is low-income and minority students who are the most likely to be subjected to impersonalized education through more affordable but highly commercialized online instruction. A real danger in the United States is that one group of students, disproportionately wealthy, may attend small-class seminars in liberal arts colleges, while another group of students, disproportionately poor, may receive an undergraduate education through online diploma mills (Noble 1998a; 1998b).

Similarly, in Africa, Latin America, and Asia, good-quality local universities could face increased competition from online programs offered by American or European universities with a standardized international curriculum and big-name professors (whose lectures are offered via video tape), but without the kind of interaction, mentoring, guidance, and locally relevant curriculum necessary to support meaningful education. Some universities in developing countries, such as the University of South Africa, are making successful use of the Internet in education (Heydenrych 2001), but in these cases the Internet serves as a supplement to other media of instruction rather than as a replacement for them.

Distance education does offer important new possibilities for learners, such as the opportunity to access a wide array of instructional programs in one's own home. However, as I have argued, distance education also generatés important challenges for educators. High-quality distance education programs demand close attention to the social context in which effective learning takes place. In many cases, that might involve a combination of face-to-face and Internet-based learning in order to maximize the advantages of both. Failure to consider the social context of distance education can actually result in worsening social stratification.

Conclusion

Human resources are one of the most important factors affecting social inclusion and exclusion. Literacy and education can be furthered through the use of technology but not merely through the provision of the hardware, software, and connections. A computer program or Web site can provide information but it cannot provide the kinds of social interaction that are at the heart of good education.

Effective technology-based educational programs, whether in community technology centers, schools, or universities, integrate active mastery of technology and engagement with challenging content. Student-centered projects are carried out, not as ends in themselves but as a process of apprenticeship toward relevant ends. Careful attention is paid to creating the social networks of interaction, networking, and support that allow learning to flourish.

In summary, information and communication technologies intersect with the struggle for better education, and not always in ways that benefit marginalized learners. The deployment of technology toward greater equality, inclusion, and access is in no way guaranteed but will depend in large part on the mobilization of learners, educators, and communities to demand that technology be used in ways that serve their interests.

6

Social Resources: Communities and Institutions

The importance of social relations in shaping access to technology has been a major theme of this book. In this chapter I deepen this discussion by focusing on the concepts, research, and practice related to the intersection of information and communication technology (ICT), community development, and institutional reform. An underpinning for this discussion is the notion of social capital.

Social Capital

The concept of social capital arose in the 1980s as a number of social scientists considered the role of interpersonal relations in human and social development (e.g., Bourdieu 1986; Coleman 1988). To many, it was clear that the long-existing concepts of human capital (individual skills, knowledge, and attitudes) and physical capital (financial assets) did not fully describe the developmental resources available to people and societies. Parallel to human capital and physical capital is a category of social relations and trust that has come to be called social capital. Woolcock (1998), for example, points to his discussions with rural villagers in India in defining the concept:

When asked to explain why such miserable conditions prevail in their village and what they think needs to be done to improve things, the villagers' answers are revealing. The main problems, they say, are that most people simply cannot be trusted, that local landlords exploit every opportunity to impose crushing rates of interest on loans, and pay wages so low that any personal advancement is rendered virtually impossible. There are schools and health clinics in the village, they lament, but teachers and doctors regularly fail to show up for work. Funds allocated to well-intentioned government programs are siphoned off by local elites. Police torture innocent villagers suspected of smuggling. Husbands regularly beat or abandon their wives. Utter destitution is only a minor calamity away. You venture that surely everyone would all be better off if they worked together to begin addressing some of these basic concerns. "Perhaps," they respond, "but any such efforts seem always to come to naught. Development workers are no different: just last month, someone who claimed to be from a reputable organization helped us start savings and credit groups, only to vanish, absconding with all our hard-earned money. Why should we trust you? Why should we trust anyone? (152–153)

In the eyes of social scientists, what this village lacks is social capital. Social capital can be defined as the capacity of individuals to accrue benefits by dint of their personal relationships and memberships in particular social networks and structures. For example, if a friend provides information about a possible job, that represents social capital. If a parent offers high educational expectations, opportunities, and support to a child, that represents social capital. If a government bureaucrat can be trusted to do what he or she says, that too is a form of social capital. Social capital accrues both to individuals and to communities, which benefit from the collective social capital in their midst. Even a poorly connected person benefits from living in a well-connected community; for example, if members of a community are known for keeping an eye on each others' homes, that will discourage crime in the neighborhood and benefit even those residents who have few neighborhood ties.

Social capital is not just an input into human development but a "shift factor" affecting other inputs (Serageldin and Grootaert 2000, 54), because it tends to enhance the benefits of investment in human and physical capital (Putnam 1993). For example, investments in training can be multiplied by the inputs of social capital as the strengthening of social ties enables people to better learn from others.

An important source of social capital is the personal relations that people have in their family and community. These relations can provide information, influence, social credentials, and reinforcement (Lin 2001). Information can include everything from a recommended health care provider, to a tip for a job opening, to advice on preparation of soil. Influence is exerted on others, for example, when an associate persuades somebody to hire you. Social credentials refer to the higher regard that someone might have for you because of your social connections (e.g., your family, friends, neighborhood). Reinforcement refers to the emotional and personal support you get from people you know (e.g., encouragement in the face of illness). Norms refer to the general expectations of the groups around you; for example, a child benefits greatly if he or she attends a school where everybody is expected to attend college.

These benefits can be shared through bonding social capital and bridging social capital (Putnam 2000). Bonding social capital refers to the strong ties that are shared among dense, inward-looking social networks, such as among family members, close friends, church groups, or ethnic fraternal organizations. These strong ties provide the kind of emotional support that allows us to get by. Bonding social capital plays a dual role: it brings the strength of social solidarity but sometimes at the cost of antagonism with or distance from other groups (think, for example, of the strong bonding social capital in a youth gang, which might serve to alienate or isolate gang members from access to other social sources of information and support).

In contrast, bridging social capital refers to the ties that are formed with those from other social circles. Since it provides important links to new sources of information and support, bridging social capital is considered especially important for economic and social development. The value of bridging social capital is explained by Granovetter's (1973) theory of the strength of weak ties. Those in our own immediate circle—our strong ties—tend to have similar friends and similar sources of information to us. Therefore, when we bond with them, we may not gain much in terms of new sources of information or support. However, distant acquaintances and contacts will have access to different people, different information, and different social networks. Therefore, a broad network of weak ties is actually more important than a small network of strong ties in many ways. For example, distant, weak social ties may be more useful than strong, close ties in finding a job or gaining political allies. As Xavier de Souza Briggs puts it, strong ties are good for "getting by," while weak ties are crucial for "getting ahead" (quoted in Putnam 1993, 21).

Other sources of social capital are the norms, rules, and expectations that exist in a neighborhood, community, or society. If men are expected to treat women well, individual men, women, and children will all benefit. If drivers are expected to stay in their own road lane and signal before turning, each driver will benefit because of the increased safety—or at the very least, predictability—of one's own and others' driving. If government officials must follow rules that restrict opportunities for corruption, the entire society will benefit. Of course *relational* social capital (based on bonding and bridging) affects *norm* social capital (based on norms and expectations), and vice versa. The types of bonds and bridges that exist between individuals, the types of groups and organizations that people belong to, and the way that these groups express the needs and desires of a community all affect and are affected by the norms and expectations of a society.

The Internet and Social Capital

What, then, is the relationship between the Internet and social capital? On the one hand, social capital is an important factor in gaining access to computers and the Internet. Entering the world of computing is quite complex. It involves making decisions about whether to buy a computer, what kind of computer to buy, how to set it up, what kind of software to get, how to install it, how to obtain and set up Internet access, and then how to use the computer, the software, and the Internet. Most people rely on their social networks to offer support and assistance in this. That might involve anything from observing computer use at a friend's house, hearing how a neighbor uses the Internet, asking a colleague to help solve a software problem, or simply buying a computer for your child because it is a general expectation in your community that children should have access to computers (see Agre 1997).

For people whose social network does not include computer users, the challenges of purchasing, setting up, and learning to use a computer can be overwhelming. Two recent studies conducted in California provide evidence of the value of social capital in gaining entry to the world of computing. The first study, based on a survey of 1,000 people, found that social contact with other computer users was a key factor correlated with computer access (SDRTA 2001). As the study reports,

Although most respondents stated that they know people who used computers, the digitally detached (those who do not have home personal computers, Internet access, or access to the Internet outside of the home) did not. And when compared with the impact of ethnicity, income, and education level, this sentiment—that they did not know others who used computers—is far more significant. (12)

The second study, based on interviews at community technology centers, found that the social support offered at those centers was critical to many people's decisions to purchase computers (Stanley 2001). In many other cases, though people already had purchased computers, they were not yet using them but began to use them after engaging in computer use in a supportive social environment offered by their local community technology center.

Neither of these studies proves that a particular type of social support promotes computer access; they do, however, show that social networks and computer usage are inextricably linked. Community initiatives can take advantage of this linkage to facilitate home computing. For example, one neighborhood project in Massachusetts—in spite of offering a package including free home computers, free Internet access, and free training—was only able to sign up 8 of 47 families in a neighborhood housing project. However, after residents from the housing project who had already participated in the first round of the project went door-to-door and spoke of the personal benefits that they had received, the registration rate jumped from 17% to 57%.[1] Similarly, a "learn-and-earn" project in Riverside, which allowed people to purchase computers at a discounted price if they had first completed a computer course at a training center, found that the personal contacts established at the training center were critical to people's use of their new home computers. It was not only the training and skills the center provided but also the network of support; the new computer owners would return frequently to the center to consult with staff members about hardware, software, and other issues related to computer and Internet use.[2]

The larger question is not whether social capital provides support for using the Internet but whether using the Internet extends people's social capital. The natural assumption is that the answer is yes, because the Internet provides expanded opportunities for communication and association with broad numbers of people. This is especially important for developing weak social ties, for which the Internet is a natural medium. As Collier (1998) explains, one of the simplest ways to promote social capital is to lower the cost of social interaction, and the Internet certainly achieves that. One leading sociologist has gone so far as to proclaim that the rise of the Internet has brought about "a revolutionary growth of social capital" (Lin 2001, 237).

Empirical studies do suggest that the Internet can promote social capital. An in-depth study was conducted by Keith Hampton, who carried out his dissertation research in a suburban housing community in Toronto known as Netville (Hampton 2001a). All those purchasing homes in Netville were offered free broadband Internet access, but in the end this access was provided to only 60% of the residents. The resulting dichotomy between those with and without Internet access—two groups that were highly similar in most other ways—provided a fertile laboratory for analyzing the impact of Internet use on social capital. The study found that those with Internet access maintained and developed more extensive social networks of contact and support both within Netville and outside (Hampton and Wellman 2001). Outside the community, the wired residents tended to maintain or increase their contacts and support from people who lived less than 50 kilometers away, between 50 and 500 kilometers, or more than 500 kilometers, whereas the unwired residents faced decreased contact or support at all three distances (presumably because they had just moved to a new neighborhood and thus were removed from old contacts and busy getting moved in to their new homes). Interestingly, the greatest differences in support rendered to wired and nonwired residents was from people at an intermediate distance (50–500 km), suggesting that the Internet is especially helpful at building social capital with those people who are "just out of reach" (more so than with those whom one never sees, or sees often).

As for contacts *within* the Netville neighborhood, those were also (Hampton 2000; 2001b) bolstered by online communication. This was due not only to Internet connectivity but also to the use of a community e-mail list known as Net-l. Because of communication on this list and the social ties that arose from exchanges online and later offline, wired residents had substantially more contact of every sort within the community than did nonwired residents, whether measured by number of people recognized by name, number of people talked with on a regular basis, number of people called on the phone, or number of people visited at home. Interestingly, the wired residents even had more contact with the nonwired residents than the latter group had among themselves, apparently because the wired residents took responsibility for sharing and passing on information from the Net-l list to their nonwired neighbors.

However, there are countervailing factors involved in any consideration of social capital and ICT use, including several reasons why the Internet might *not* promote social capital (see Putnam 2000). First, as discussed in chapter 5, face-to-face interaction provides a richer form of communication and support than does online interaction. To the extent that online communication *supplants* rather than *supplements* face-to-face interaction, it could thus weaken social capital. Think, for example, of a school class that carries out an international exchange with students in another country while missing opportunities to interact more directly with different social or ethnic groups in its own city. At least one study claims to show that the more time people spend online a week, the more they lose contact with their social environment (Nie and Erbring 2000).[3]

This potentially negative effect on social capital could be exacerbated by the amount of hostility that occurs online. The reduced communicative content (no visual or auditory clues) frees people up from their inhibitions online, which allows easy contact with large numbers of people but also can bring out the worst in people. This results in a phenomenon called flaming, in which people express hostility in ways they might never do face-to-face.

The Internet can also lead to a narrowing of social contact rather than a broadening. A teenager is just as likely to spend hours online chatting with a small circle of friends as he or she is to form new contacts and bridges with diverse social networks. Those who use the Internet to seek information may also have their sources narrowed rather than broadened (see Sunstein 2001). The Internet continues a trend of narrowcasting that began with the proliferation of radio stations and television channels. On the Internet, you can design "*my* CNN" or "*my* Yahoo," thus making it less likely that you would discover the new sources of information that you might come across in reading the newspaper or browsing a library shelf.

Finally, there is no assurance that people will use the Internet for either social interaction or information. The most popular and fastest growing uses of the Internet include private, antisocial forms of entertainment, such as viewing pornographic material and gambling. To the extent that the Internet facilitates activities such as these, it will weaken rather than strengthen social capital.

While some cyber-pessimists have sounded the alarm about these potential drawbacks of the Internet, most sociologists take a more balanced view. The associative power of the Internet can be exploited to supplement social capital, but not if the Internet is seen as the be-all and end-all. Rather, strategies must be devised to combine the strengths of the Internet with other forms of interaction. This is especially important when working with impoverished or marginalized groups that need to leverage *all* their sources of capital in order to thrive.

How can this be accomplished? Efforts to make use of ICT to promote social capital take place at three different, albeit overlapping, levels.[4] One is the microlevel, referring to the relations with friends, relatives, neighbors, and colleagues who provide companionship, emotional support, goods and services, information, a sense of belonging, and opportunities for community development. A second is the macrolevel, which corresponds to the effectiveness of governmental institutions and transparent and trustworthy relationships that exist between governments and citizens. A third level, falling between these two, is the mesolevel, corresponding to the voluntary associations and political organizations that allow people opportunities to form alliances, create joint accomplishments, and collectively defend their interests.

Microlevel Social Capital: Virtual Community vs. Community Informatics

There are two approaches to using the Internet to promote microlevel social capital. The first can be called the virtual community approach, and the second, community informatics.

Virtual Community
The term *virtual community* was popularized by Howard Rheingold in 1993 with the publication of *The Virtual Community: Homesteading on*

the Electronic Frontier. Rheingold had been active in computer conferencing since the mid-1980s, and he wrote eloquently of people's experiences in a pioneer computer conference called the WELL (Whole Earth 'Lectronic Link). The WELL's members, mostly upper-middle-class suburbanites in the San Francisco bay area, participated in a wide range of electronic forums such as Arts and Letters (the Beatles, Jazz); Recreation (Gardening, Chess); Entertainment (Star Trek, Bay Area Tonight); Education and Planning (Biosphere II, Transportation), Computers (Software Support, Desktop Publishing), and Body-Mind-Health (Recovery, Gay Issues). Rheingold provides a compelling narrative of how a group of strangers from different backgrounds and places came together electronically to share information, debate and discuss ideas, and provide emotional support in time of need.

However, Rheingold later came under broad criticism for his overenthusiastic reliance on anecdotal evidence, his neglect of countervailing patterns, and his apparent support for the notion that online communities were distinct from, and perhaps better than, traditional ones. To be fair, he never claimed to be carrying out academic research, and his views on the existence of an autonomous cyberspace and virtual communities were never as extreme as those of others, such as cyber guru John Perry Barlow of the Electronic Frontier Foundation, who argues that cyberspace constitutes an entirely different world (Barlow 1996).

Eventually, Rheingold himself came to moderate his views, partly because of his participation in bitter social conflicts that emerged in and around the WELL. In the revised edition of his book (2000), he questions the very notion of virtual communities as distinct from traditional communities. His critique rests on two key points. First, any technology emerges from and responds to existing social relations and social contexts. Technologies may create new possibilities, but they do not in themselves represent separate worlds. Second, and related to the first point, the division between virtual and traditional communities is spurious. As social network theory and research make clear (e.g., Wellman et al. 2001), so-called traditional communities are rarely based on neighborhood or village alone. In all but the most isolated parts of the world, people's social networks include relatives, friends, and other associates who live elsewhere, and contact with them is maintained through personal visits, mail, telephone, or other media. Thus the notion of neighborhood-based traditional communities is outmoded to begin with.

Extensive research in a variety of domains indicates that the use of ICT tends to complement rather than replace other means of networking (Hampton and Wellman 1999; Warschauer 1999; Wellman et al. 2001). One of the best discussions of this issue is Philip Agre's (2001b) primer for graduate students, "Networking on the Network." Agre explains how students can make use of the Internet to steadily increase their contacts and positioning within academia, but only through building on other kinds of institutional relations already existing in academia, such as personal contacts between students and professors at universities and conferences.

Because of the intersection between online communication and other means of social networking—and because of the limitations of online interaction discussed earlier—approaches to promoting social inclusion that rely solely on virtual communities are ill-advised (see, for example, the discussion on distance education in chapter 5). Successful approaches usually combine online and face-to-face networking. An example of this is a group of Egyptian educators who have successfully come together (outside Ministry of Education channels) to learn about educational technology and its integration into their classrooms and lives. The group thrives through an e-mail discussion list, but this list was only formed after several face-to-face meetings in Cairo to share ideas and discuss plans. The group still holds occasional face-to-face meetings to conduct training sessions, hear guest speakers, or simply socialize. Annual technology fairs are planned for national conferences. Through a combination of face-to-face training, committees, projects *and* e-mail discussion, the group has developed into a strong educational support network—far stronger than it could have become through Internet contact alone.

Community Informatics

If the virtual community concept provides a poor frame of reference for thinking about technology for social inclusion, a better framework is provided by community informatics (see Gurstein 2000; Loader et al. 2001). Community informatics seeks to apply ICT to help achieve the social, economic, political, or cultural goals of communities. Community informatics begins from the perspective that ICT can provide a set of resources and tools that individuals and communities can use, initially to provide access to information management and processing and eventually to help individuals and communities pursue goals in local economic development, cultural affairs, civic activism, and community-based initiatives (Gurstein 2000). Community informatics strives to take into account the design of the social system and culture within which the technology resides as well as the design of the broader technological system within which a particular tool or medium interacts.

Promoting social capital is a key strategy of community informatics, but this is not seen as taking place principally through online communication. Rather, social capital is created and leveraged by building the strongest possible coalitions and networks in support of the community's goals, using technology projects as a focal point and organizing tool. Online communication is of course part of this, but so are more traditional forms of communication, organization, mobilization, and coalition building.

My research in a variety of settings has indicated that five strategies are critical to promoting social capital in community technology projects. These include leveraging existing community resources, mapping and connecting existing community connections, integrating with broader social and economic campaigns, organizing new social alliances, and social mobilization via a wide range of media and tools.

Leveraging Community Resources   Probably the most effective method for leveraging community resources is to work through existing community organizations or leaders to launch and manage community technology initiatives. Community organizations know the local situation and can work to provide a structure that meets local needs. One of the main failings of the Hole-in-the-Wall computer kiosks in New Delhi discussed in the introduction to this book was that no local community organization was involved in running them. The director of the project, a government official, told me that he preferred to work "directly with the people" rather than via intermediaries such as community organizations.[5] However, in this case, going "directly to the people" meant placing a project in a community without any organized way for the community to partake in the management or leadership of the project. In contrast, the best community projects I have witnessed in India, Brazil, and elsewhere make ample use of community resources. In India, for example, where many local villages are led by local tribal leaders, their participation is critical in the effective implementation and management of community technology projects. In the cities of Brazil, in contrast, key community roles are played by neighborhood associations, slum-dweller groups, and other nongovernmental associations.

Many times, a single community group can be brought in directly to manage a project. In other cases, leaders of several community organizations can come together to form a new community council for the project. So while the Hole-in-the-Wall project suffered from an unclear purpose and questionable sustainability because of its lack of community leadership, another street children's project in India—a computer-based training program—was run very effectively by a nongovernmental organization called Prayas, which has a long history working with street children in many other capacities (e.g., through housing programs, health clinics, counseling services). Since this group already has worked with street children, knows their needs, and has their trust, it was able to fashion a computer training program that had a better organization (through structured classes) and purpose (vocational training) than merely placing computers in a slum.

Another strategy for leveraging community resources is through the promotion of ICT capabilities of extant groups. An example of this is the Community Digital Initiative, a community technology center in Riverside, California, a city with a large number of low-income Latinos.[6] Though the center also runs individual training programs, much of its impact comes from work with organizations. The project chose as its location a large building in the center of town that hosts many other community organizations, including violence prevention groups, a dispute mediation group, a volunteer center, court referral programs, housing programs, transportation access programs, crisis intervention programs, community health projects, and legal aid organizations. These groups, as well as similar organizations elsewhere in Riverside, form an important part of the clientele of the center. Managers, administrators, and members of these groups participate in workshops to learn how to use computers and the Internet to function more effectively. This might include everything from developing a computerized mailing list, keeping track of organizational finances, developing brochures and newsletters, making an organizational Web site, or setting up an internal or external e-mail list. The center also provides its equipment and resources for the community organizations to carry out these tasks. In this way, the efforts of the Community Digital Initiative are multiplied as existing community groups are empowered to carry out their organizing more effectively.

Mapping and Connecting Community Resources   Mapping community resources is a critical component of launching a successful community technology initiative. Participatory rapid appraisal (PRA) techniques (see chapter 4) can be used to identify relations and resources within a community and how those might be amplified through a technology initiative. For example, one of the PRA techniques used in rural tele-center planning in India is community mapping. Different members of the community are invited to draw maps of the community from their own perspective. By seeing how different members map the community, organizers can learn about which locations, people, and assets of the community are most valued, and which location, for example, might be the best for placing a community telecenter. Other PRA techniques include identifying who is in contact with whom. For example, another rural Indian telecommunications project learned through PRA techniques that women in the village tended to communicate principally with other women rather than with men. The project then incorporated the rule that half the staff of its village knowledge centers should be women, to ensure that everyone in the village would be able to have a voice and be heard.[7]

As discussed in chapter 4, the community can later be drawn into developing more detailed databases of local resources and assets. For example, the Sampa.org project in São Paulo has mobilized a group of people to develop a geo reference system of the Capo Redondo *favela*, including a physical map of the streets and a database of community resources such as health centers, neighborhood associations, and worker cooperatives. Whereas this kind of information is easier to find in wealthier communities, many of the *favelas* of Brazil are like informational black holes. In this case, a commercial transportation company, which monitors deliveries in the area but has no existing map, funded the project. From the community organizers' point of view, the project serves multiple ends. Not only will the database itself be of great use to community members when made available through the local telecenters in the Sampa.org project, but the mapping team is gaining important vocational skills (e.g., development of geo reference systems) that they can then market to others.

Integrating with Broader Social and Economic Campaigns   Technology projects do not exist as ends in themselves. They are most effective when they are tied to broader social and economic campaigns, as seen by the following three examples.

Bresee Cyberhood   The Bresee Foundation has been carrying out community organizing in central Los Angeles since the mid-1980s.[8] The neighborhood in which it organizes is one of the poorest and most crime-infested neighborhoods in Los Angeles, with high rates of homicide, gang-related shootings, auto theft, domestic violence, and drugs, and low rates of employment, income, and health care. More than 40% of the people in the neighborhood live below the poverty line and some 54% lack health insurance. The neighborhood was also the site of the infamous Rampart police scandal, in which members of the Los Angeles Police Department were convicted of planting evidence, faking confrontations, and repeatedly lying to send men to prison.

Bresee's community development strategy is largely built around forging social capital in the neighborhood. This is accomplished via a community center that provides a safe, trusting environment through a wide array of programs. The center includes a health clinic, a homework assistance program, a recreation program, youth discussion groups, and employment training. Recently, Bresee opened a computer center, known as Cyberhood, within the broader community project. Cyberhood offers a wide range of services, including open drop-in computer and Internet access for adults, a range of computer courses, a multimedia internship program for teenagers, and technology-and-employment programs.

The integration of Cyberhood into the broader social development mission of Bresee offers many benefits. For one, Cyberhood builds on the safe and trusting atmosphere of its parent center. Among the Cyberhood participants I interviewed was a local homeless youth who came to the Bresee center regularly because it offered a safe environment for him to spend time in. Also, the computer projects can benefit from the broader organizing campaigns and relationships of Bresee. Bresee's relations with local schools, universities, and businesses, for example, help to recruit the right children into its technology programs and to provide them with sufficient volunteer support. In addition, there is much crossover between the different services. Some of the people who attend the health clinic at the center later come to Cyberhood to find information on the Internet related to their health needs.

One strategy for community building is through fostering local leaders. Some 25% of the employees at Bresee were formerly clients of the center. Local youth leaders are constantly developed through training and internship. For example, in the computer center, much of the hands-on assistance is done by teenagers who have completed a course there and who have demonstrated talent; they are subsequently hired to help others.

The Cyberhood programs also serve the economic goals of Bresee, through integration of technology and employment training. Courses focus on graphic business design skills, such as the development of business cards, brochures, and newsletters. Other software programs are used to help the youth develop an overall sense of direction in their lives. For example, a software program called Choices helps people research the kinds of careers they might be interested in and the types of preparation those careers require.[9]

In summary, Cyberhood is not an end in itself. As the director of Bresee explained to me,

We don't just teach people computers, it's not just about developing skills—it's about connections with people and building relations. This community lacks the kind of mediating institutions like good schools, churches, and parents involved in the schooling. Our technology programs work together with all our other programs to help people develop these kinds of relations that are often missing. In this way we can be a gathering place and hub for the community.[10]

M. S. Swaminathan Research Foundation   A rural counterpart of Bresee is M. S. Swaminathan Research Foundation (MSSRF) in southern India.[11] MSSRF has been carrying out economic and environmental programs in communities in Pondicherry and Tamil Nadu since 1991. MSSRF works with the neediest groups in order to simultaneously combat rural poverty and environmental degradation. Its strategy in rural India is to help landless laborers and small farmers develop the skills, resources, and organization they need in order to obtain much greater value from their labor. As a centerpiece to this, they have developed two model biovillages, where agricultural laborers can come to observe environmentally sustainable farming processes firsthand and learn new skills, techniques, and information. Projects at the biovillages focus on aquaculture, mushroom and flower cultivation, fodder cultivation, horticulture, conservation of rainwater, composting, rope-making from coconuts, pest control, and dairy farming.

MSSRF later developed village knowledge centers, a network of computer kiosks in rural villages, to serve this broader socioeconomic development project. Content from the biovillage projects is made available throughout an intranet that connects the centers. Even if the farmers themselves can't read it, the center staff can share information about bio-farming with them. With funding from the Commonwealth of Learning, a local farmers' group is further developing this content into databases to assist rural development campaigns throughout India. In addition, MSSRF is helping women's collectives learn computer skills needed for microfinance management, so that they can better work to obtain and manage their bank credit in carrying out sustainable agriculture projects.

One of the more exciting offshoots of this program is Oddanchatram-market.com, an e-commerce Web site and campaign started by a local farmers' association. In order to enhance demand—farms in the area lay fallow 40% of the year because of lack of a market for the goods—the local small farmers' association went to the suppliers and offered to announce their goods on a Web site. The intention is to increase national demand for local products, thus providing greater income for suppliers, farmers, and agricultural laborers alike. The suppliers will pay a nominal fee for the service, thus providing additional funds for the farmers' association.

ISIS for the Blind and Visually Impaired   The Information, Service, Integration, and Schooling (ISIS) project for the blind and visually impaired is located in Graz, Austria.[12] ISIS, whose president and staff are themselves all blind or visually impaired, seeks to open job possibilities beyond classical blind employment (e.g., basket weaving) to more wide-ranging opportunities in the information economy. Toward that end, ISIS has set up a computer training center offering courses for the blind, ranging from an introductory module leading to a European Computer Driving License[13] to specialized topics like the Linux operating system, with an orientation toward labor market needs. The training center also hosts what is believed to be the first public Internet café for blind people, where people are offered the opportunity to browse the Internet using specially designed hardware and software as well as to interact with other café visitors. In addition, a telephone "blind line" provides free information to the blind and their relatives, educational establishments, and social service agencies about assistive technology issues and training opportunities.

Organizing New Social Alliances   While the virtual community approach focuses on developing online ties, the community informatics approach seeks to actively engage an array of groups in social projects. In this way, the community gains access to social contacts and support from diverse resources that may not have been accessible before. For example, Alkalimat and Williams (2001) describe how gradually increasing the involvement of local churches, a university, and municipal organizations provides strength and sustenance to a community technology center. Similarly, in Egypt, the 21st Century Clubs are a national set of computer centers that have been developed through an alliance of nongovernmental organizations (which run the centers), private businesses (which donate the computers), software companies (which provide office and edutainment packages), and the governmental Ministry of Information Technology (which coordinates the project).

The Riverside Cybrary in California is an example of an initiative that helps develop social capital, in part through strengthening community-university relations. The Cybrary was initiated by the Riverside Public Library. It is located in a storefront in a low-income Latino community and provides a friendly, accessible atmosphere for neighborhood youth. Children and teenagers drop in to use the computers in the library, all of which are connected to the Internet, and are offered individual instruction and support from local volunteers. Most of the volunteers come from special service learning projects run by local universities. The volunteers not only provide invaluable computer support but also act as role models for the teenagers, many of whom have never met anybody who has attended college. Since many of the volunteers are themselves Latinos, they can relate well to the children and answer their questions not only about computers but also about college and possible career paths.

In regard to outside support, it is important to bear in mind that no group participates in a community project without its own interests in mind. For example, many community technology projects are aided by businesses that donate hardware or software or provide community volunteers, and one of the largest such supporters is Microsoft Corporation. This support often comes in the form of donations of Microsoft software. While free software is valued by projects, these donations also benefit Microsoft in that they commit projects to a Windows platform and showcase Microsoft products to potential future customers. In addition, community organizations are much less likely to pursue using free software solutions, such as those based on Linux, if commercial operating systems and office suites are offered free.

Social Mobilization through a Variety of Media   Another important strategy for community technology development is to use all available media to amplify the power of the Internet. This is especially important in developing countries, where individual use of the Internet is not widespread. Examples of this principle are three projects in South Asia.

The Kothmale Community Radio Internet project in Sri Lanka makes use of FM broadcasting to bring online information to thousands of people without Internet access.[14] Kothmale is located in an underdeveloped area of Sri Lanka, several hours from the capital. The radio station was set up in 1989 to serve the needs of the rural and small-town population, numbering close to 350,000 people in a twenty-five-mile radius. An Internet component to the project was launched in 1998 with a grant from UNESCO. The Kothmale radio announcers gather information from the Internet, which they incorporate into news, weather, journalism, and music programs. The announcers also take questions from listeners delivered by postal mail, research the questions on the Internet, and provide answers on the air in local languages. The questions are answered on radio programs focusing on topics such as human rights, the status of women, rural health, farming, and international events. The questions are often very specific to the needs of the community, such as the care of a local tropical disease or the best way to raise and market geese. The announcers also visit the rural villages themselves to get local content for their broadcasts as well as to gather more questions from the community. Finally, to provide greater Internet access, the radio station has opened up its own facilities for community Internet use and has also built two more community technology centers in the area.

The M. S. Swaminathan project, mentioned earlier, makes very creative use of multimedia in one of the village knowledge centers located in a fishing village. The project staff members download information on weather and sea conditions daily from the U.S.-based Cable News Network and U.S. Naval Station Web sites. They translate the information into the local Tamil language and broadcast it over a loudspeaker from the community technology center. The village fisherman can easily hear the broadcast from the beach 100 meters away, and they use the information to improve both their safety and productivity. Information is also posted on blackboards and bulletin boards outside the center for those who might walk by.

Finally, the Gyandoot rural technology project in Dhar makes use of a wide variety of media and campaigns to build involvement in the project while focusing attention on local health and economic concerns. One of its annual campaigns is a "healthiest child competition," in which parents throughout the rural area are invited to bring their children to the local village Internet kiosk. Radio and newspaper announcements, posters, and even local theater boards are used to promote the competition. Volunteer medical personnel come to the Internet centers to weigh the children, check their vaccination records, and otherwise perform examinations. The campaign serves to introduce the rural populace to the Internet kiosks while also calling attention to important health needs, such as vaccinating children. A similar campaign was later organized for the "most productive cow," based on milk yields and promoting knowledge of dairy farming techniques.

In summary, all the examples discussed in this section use technology as an additional tool to promote social capital and community development, and none of them focus on technology as an end in itself. Information from the Internet is used to enhance this process, but it is often downloaded and shared via a variety of media rather than expecting individual use by community members.

Computer-mediated communication can be an element of community informatics, but it is not the only or even the principal form of communication. Rather, as Resnick (2002) suggests, it is more likely that an Internet-based communications system can be used effectively for community development only after its users have developed, through other means, trust in each other, a shared identity, or some other form of social capital.

Macrolevel Social Capital: Governance and Democracy

If microlevel social capital comes from the bottom up, macrolevel social capital comes from the top down. It concerns how the social structures of large institutions, especially governments, provide and facilitate resources and support to individuals and society. In this section I examine issues of governance and democracy, and their relationship to technology and social inclusion.

As Woolcock (1998) explains, an important component of macrolevel social capital is synergy, in other words, congruent and positive relations between the state and society. Woolcock cites India as a prime example of a country that, although democratic politically, and with a well-educated and highly prestigious civil service, still suffers from a severe lack of synergy. In India and other "weak" states,

The government may be committed in principle to upholding common law and may refrain from actively plundering the common weal, but in practice misappropriates scarce resources, is largely indifferent to the plight of vulnerable groups (women, the elderly, poor, and disabled), produces shoddy goods, responds slowly if at all to citizen demands, and is notoriously inept in supporting businesses seeking to be competitive in world markets. (177–178)

This lack of synergy worsens divides in India and other low-income countries in two ways. First, it holds down national development and keeps these countries poor vis-à-vis the West. Second, it serves to maintain and increase inequality within the country because the most marginalized and vulnerable members of society suffer the most from lack of governmental support.

Developing synergy is a challenging task, especially in countries with high degrees of inequality. A vicious circle often develops, in which the marginalization of the poor (through lack of literacy, social isolation, lack of access to media) puts them at a distance from government officials, governmental information, and governmental programs. This lack of access to governmental resources serves to increase their poverty and marginalization, which in turn further weakens their access to governmental assistance.

Well-designed use of ICT can help break this pattern and replace it with a virtuous circle of increased access to governmental information and resources, less marginalization, and further increased access. For this to take place, e-governance programs have to be carefully designed with the needs of the poor and marginalized in mind; otherwise, such programs will likely only benefit those who are already well connected. E-governance initiatives can help the poor in at least two ways. First, they can make government *transparent*. Second, they can help facilitate *citizen feedback*. I discuss initiatives in each of these areas using examples from India.

Transparency

In a typical developing country, gaining access to even the most common types of governmental information or documents can be a nightmare. Obtaining a simple governmental record can involve one or more overnight bus trips to the state or national capital; waiting in long lines in hot, overcrowded, and poorly organized government buildings; shuffling back and forth between a host of departments; and, too often, paying a hefty bribe to eventually get the document needed. And whereas the well-to-do often assign their servants to carry out such tasks, the poor have no choice but to carry out such tasks themselves, losing a good chunk of their meager income on transportation, lost wages, and bribes.

In many cases, the obstacles are so overwhelming that the poor don't bother even to try seeking their rightful information or documentation, and over time the lack of information and documents serves to worsen people's economic and social marginalization.

Developing transparent information and documentation systems is no easy task, especially given the low wages of government employees in most developing countries. Those who would be most responsible for implementing more transparent systems have little incentive to do so because that would lessen the possibilities for bribes and thus decrease their own incomes. It thus becomes almost impossible to achieve transparency through moral appeals to individual employees or units to improve the quality of their work. Rather, systemic solutions must be developed at a broader and more comprehensive level. The use of ICT to systematize the maintenance and distribution of governmental information and documentation provides one possible mechanism for achieving transparency, especially when combined with efforts to make sure that the marginalized have equal access to this computerized information.

The lack of systematic and transparent recording and public documentation of governmental data is a major issue affecting international development. Probably the most important example of this is in land records. Hernando de Soto (2000), an internationally renowned economist and president of the Institute for Liberty and Democracy in Peru, has published an exhaustive and compelling study of the importance of land records for international development. De Soto explains that, unlike the West, where transparent documentation of land ownership allows people to use their property as a source of capital, much of the developing world lacks such documentation, with a devastating effect. He writes,

In every country we researched, we found that some 80% of land parcels were not protected by up-to-date records or held by legally accountable owners. Nobody can identify who owns what, addresses cannot be easily verified, people cannot be made to pay their debts, resources cannot conveniently be turned into money, ownership cannot be divided into shares, descriptions of assets are not standardized and cannot be easily compared, and the rules that govern property vary from neighborhood to neighborhood or even from street to street. (Quoted in Binns 2001, 2)

According to his research, the total value of the real estate held but not "legally" owned by the poor of developing countries and former communist nations is equal to some 9.3 trillion USD. In most such countries the value of this extralegal real estate is many times greater than total savings and time deposits in commercial banks, the value of companies registered in local stock exchanges, all foreign direct investment, and all privatized public enterprises. If even a small portion of this amount of capital were unleashed, it could present an enormous reservoir for economic development and poverty alleviation.

India provides an excellent example of both the importance of land records and the difficulty of obtaining them.[15] Government land records in India contain an exhaustive amount of information, including a delineation of the property borders, a list of crops grown, a description of crop output, a list of the cultivators and tenants, and a report on any outstanding agricultural loans from government agencies. Copies of land records are required for a wide variety of transactions, including long-term land mortgages, short-term crop loans, and applications for government poverty alleviation programs (e.g., to demonstrate that the person is a small farmer), and are even used in criminal proceedings (e.g., to give assurance that the accused is a landowner and thus has geographical roots and economic means). Since these records are frequently updated, simply having one permanent copy is not sufficient. Rather, people need to go to government offices to get an up-to-date copy of the land record on most of the occasions that they need to show it.

There are hundreds of millions of these land records in India. Until recently, they had all been kept on paper, much of which is yellowed, badly faded, or torn. These records are maintained by tens of thousands of local village accountants, who are responsible for recording updated information and also for distributing copies of the land records upon request from citizens. Both aspects of this process—the recording of information and the distribution of copies—are subject to a great deal of corruption. In a paper-based system it is easy for accountants to claim that a record has been misplaced, and thus to make the farmer come back again and again to receive a copy. On other occasions, information on the land record is illegible, thus requiring another bribe for the accountant's clarification or correction. Similarly, when a sale is made, the accountant can delay recording the transaction until the new owner has paid a bribe. Finally, village accountants, who have many functions in the three or four villages they are responsible for and yet who are often out of the office, can simply ask for a bribe merely to show up to meet someone. My own interviews with small farmers throughout India indicated that bribes for land records are the norm rather than the exception and that these payments can add up to as much as a small farmer's monthly income.

In response to this situation, the state government of Karnataka, India, implemented an ambitious computerization system of land record maintenance and distribution. The project was designed not only to move from paper to computer but to move from an informal, unregulated system to a transparent, efficient one. As a first step, local governments throughout Karnataka digitized the information on all twenty million land records in the state. At the time of digitization, landowners were given an opportunity to examine the computerized record to ensure that it was correct. After computerization, land records could be updated by village accountants via computer, and only after they gained access to the system through fingerprint identification. This guaranteed that changes and updates would not be lost and that an accountant could not deny that he or she had made a change.

The computerized system allows government auditors to easily verify if all land sales have been properly recorded within the mandated thirty-day period. When people need copies of land records, they are now generated and printed out by machine. The distribution process has been taken out of the hands of village accountants (some 9,000 in the state) and shifted to a single person and machine in each subdistrict office that serve no other function than full-time printing and distribution of land records. Obtaining a land record in Karnataka is now as simple as standing in a short line and paying a fee of $0.30 USD. Those people who do not need a copy of their land records, but just want to see them, can use a self-serve kiosk to view a record for a fee of $0.04 USD.

Because of the low cost of labor in India—including in the information technology industry—the cost of implementing this system was only about $5 million USD, or some $0.25 USD for each of Karnataka's 20 million land records. This amount will be recouped over time through the $0.30 USD fee for distribution of land record copies. Of course, the economic benefit to the state, through better availability of land records, will be much greater. As Jeffrey Smith (2001, para. 3) asks, quoting de Soto,

> In a "world where ownership of assets is difficult to trace and validate and is governed by no legally recognizable set of rules; where the assets' potentially useful economic attributes have not been described or organized; where they cannot be used to obtain surplus value through multiple transactions because their unfixed nature and uncertainty leave too much room for misunderstanding, faulty recollection, and reversal of agreement," how can people prosper?

In Karnataka, because of the well-designed use of ICT to promote more transparent land record documentation and access, more people will now be able to prosper.

One important question to ask at this point is why Karnataka has been able to implement such an advanced system when no other Indian state and few other developing countries have done so? This is due partly to the technological infrastructure in Karnataka; Bangalore, the state capital of Karnataka, is the most important ICT hub in India and one of the major ICT centers in the world. Beyond that, though, the new state government, which took office shortly before this new system was implemented, had a strong vision of using ICT for human and social development. This is thus another example of how technological capacity must be combined with vision, leadership, and a commitment to social development in order to achieve an impact.

Citizen Feedback

State-society synergy cannot be fully developed just by providing information and documentation from the top down. Some kind of mechanisms for communication from the bottom up must also be provided. Citizen feedback to government acts as a check on bureaucratic abuse and corruption, alerts the government to citizens' needs and concerns, and gives citizens a sense that they have a voice in society.

There are many means for giving citizens voice in government, such as providing free elections, a free press, and opportunities for organized public protest. However, all these means have their limitations. Elections take place only at intervals and are often heavily influenced by large campaign donations or bribes, or by issues of patronage. A free press is similarly shaped by financial considerations, with media outlets reflecting the views of their owners or advertisers as much as those of the public at large. Opportunities for citizen association and protest are vital, but they are often not easily used by the poor, whose time and energy are dedicated to wage earning and survival.

Many people have looked to the Internet as a means for providing more rapid and flexible feedback to their governments. The possible advantages of this must be weighed against the possible disadvantage of giving greater voice to those who are already relatively privileged. That is one of the problems, for example, with e-voting, which could skew voter turnout to those who have computers and home Internet access, and who in most countries are disproportionately among the economically well-off.[16]

E-governance can help give voice to the marginalized if projects are designed specifically to reach the poor. An excellent example of this comes from an effort in India, the Gyandoot (purveyor of knowledge) project in the Dhar region.[17] I have briefly referred to this project earlier and now explain it in more depth because it represents a fascinating example of e-governance in one of the poorer regions of the world.

Dhar is a mostly rural district in Madhya Pradesh, the second poorest state in India. About 1.7 million people live in the district, spread out over some 100 villages. The vast majority of the population in Dhar comprises small farmers and agricultural laborers, and 57% are illiterate. According to organizers of Gyandoot, some 60% of the district residents are below the poverty line, which is defined in India as lacking sufficient nourishment. Some 54% are members of tribal groups, including large numbers of low-caste members.

The Gyandoot project was initiated by the district administrative leadership to overcome poverty and social marginalization. Unlike most ICT initiatives in rural India (and other countries), it was initiated neither by foreign donors or international agencies, nor by private business, but by local government officials in an impoverished region. The goals of the Gyandoot project are to provide better governmental information and services toward enhancing economic and social development. Its initiators have targeted the poor and marginalized and have been largely successful in reaching their audience; some 87% of Gyandoot users have incomes of less than $400 per year, and 53% are members of tribes or lower castes (Rajora 2002).

Gyandoot has two main components. One component is a collection of Internet kiosks throughout the district. Some thirty-six kiosks have been set up to date, each one managed by a local entrepreneur, who works either on his own or on behalf of the local village council. The Gyandoot project supplied a computer and phone line to each of the kiosks. The managers charge a small fee for their services (usually about $0.10 USD per transaction) to offset the ongoing costs of the operation and to earn their own income. Most users who come to the kiosk do not use the computers themselves but purchase services or information via the kiosk owners. The principal exception to this is the large number of children, who are sent by their parents for individualized computer instruction from the kiosk managers.

There is little information available on the Internet of interest to the Dhar villagers, and even less in the Hindi language. Thus the second component of the Gyandoot project is a districtwide intranet of Hindi-language information that has been especially developed for the needs of the rural poor. The intranet is developed and maintained by a small team in the Dhar district government offices and includes a wide range of information: copies of land records (though not yet in a complete and updated fashion as in Karnataka), governmental forms, applications for governmental permits, information about governmental programs (especially the many Indian programs that are designed to serve tribal members and people below the poverty line), and market rates for local crops. One section of the intranet is devoted to e-education and includes sample questions for state exams, educational quizzes, mathematical puzzles, and career guidance information.

Probably the most interesting part of the Gyandoot intranet, though, is that it also allows for a two-way process of communication. Citizens cannot only receive information but can also post it. Interactive services include an online market place (where people sell cows and bicycles), an online matrimonial service, and an online complaint service. From my interviews with kiosk managers and users in several Dhar villages, I learned that the online complaint service is an especially valued component of the Gyandoot project, and one that has had an important impact on villagers' lives.

The online complaint service comprises a Web page with a pull-down menu from which users can choose from twenty-one predetermined categories, including

• Nonpayment of salary, stipend, wages
• School closed or teacher absent from school
• Absence of a veterinary doctor
• Complaint against the secretary of village council
• Nonpayment to farmers at auction centers
• Complaint against agriculture laborer accident insurance
• that Hand pump or transformer not working
• Complaint regarding beneficiary schemes for the members of tribes and lower castes

The district administration has guaranteed to respond to complaints in each of these twenty-one categories in seven days or less. This system is enforced through public posting of outstanding complaints on the intranet through the government district offices. That way, it is immediately obvious to both government employees and their supervisors which complaints have and have not been answered. Beyond the predetermined selections, people can make complaints on any issue they wish, but the government does not offer the seven-day guarantee on these.

Interviews with villagers indicated that the complaint system was highly popular and effective. The two most common issues mentioned involved hand pumps and schools. Both of these issues speak to the lack of government response that is common in rural India, and the power of more transparent, interactive communication to help improve such responsiveness.

Villagers in India get their water from wells by means of hand pumps that are typically spread out at 1–2 km distance from each other. If the nearest hand pump isn't working, villagers must make a tiring walk to the next hand pump and then carry the water all the way back to their homes. Prior to the development of the Gyandoot system, hand pumps in Dhar frequently fell into disrepair for months at a time because government officials had little incentive to maintain them. Now, for $0.20 USD, villagers can issue a public complaint about a broken hand pump and be virtually assured that it will be repaired within a week.

A second, and in the long run more important, issue for social development involves public schooling. Villagers and government officials alike in India complain that many village schools are poorly run. They may consist of a single teacher who never shows up at all or of a few teachers who show up rarely with the approval of an also-absent principal. The lack of accountability in the Dhar schools contributed to a vicious circle: the less often teachers showed up for work, the more families became discouraged and kept their children from school; the less often children came to school, the more teachers felt justified in not showing up for work. Resigned to poor schools, and lacking recourse to complain, villagers were often forced to accept the situation. With schools, as with hand pumps, villagers now feel they are in a position to defend their rights. Complaints about absent teachers or nonfunctioning schools were frequent in the early stages of Gyandoot and were replied to promptly by government officials. These complaints have now apparently slowed down as teachers and principals become more aware that their behavior is under public scrutiny.

This last point illustrates the benefit of this interactive system even for those who might never make use of it. According to informal reports, not only has schooling improved in the district but so have hand pump maintenance, provision of public benefits, and other governmental services. Basically, government officials are now aware that their performance will be subject to public complaint and criticism. Knowing that they will be held accountable for their work provides incentive for them to perform better. Well-designed use of ICT for society-government communication—as in the case of the Dhar district in India—provides a means of improving the social capital of those directly using the online service as well as the broader community.

Democracy

Another area related to macrolevel social capital and ICT access is democracy. The relationship of democracy and Internet diffusion is a broad theme that could in itself be the topic of an entire book. I restrict myself here to a few comments and focus in particular on how this theme intersects with issues of ICT access.

Not surprisingly, prior research indicates that political openness and democracy are correlated directly with the spread of associational technologies, though not necessarily of broadcast technologies. So while the diffusion of television (a broadcast technology) is fairly uniform across societies, the diffusion of the telephone—a technology that facilitates private, horizontal communication and association among citizens—is positively correlated with measures of democracy and political openness (see, for example, a study by Buchner 1988).

Similarly, Robison and Crenshaw's (2000) international study of seventy-five countries found a substantial correlation between Internet diffusion and political openness and democratization. This is evidenced by the many authoritarian countries, such as North Korea, Syria, Saudi Arabia, and Sudan, which have put limitations on the rights of Internet users and service providers. One strategy for extending Internet use would be, it seems, to work toward greater democratization of authoritarian regimes.

However, the situation is more complex than this position implies, as witnessed by countries such as China and Singapore, in which authoritarian regimes have placed restrictions on the Internet while still allowing its rapid diffusion. In these cases, the country's leaders have embarked on a path of fast-paced economic development and are determined to make use of whatever tools are needed to accomplish the task. Indeed, both China and Singapore see the technologization of society as critical to their nation's continued economic success. In cases such as these, the more pertinent question is not whether democracy makes Internet diffusion possible but whether the spread of the Internet helps bring about democratization.

The short-term answer to this question is obviously no. Neither China nor Singapore have suddenly turned into Jeffersonian democracies simply by virtue of the fact of having large numbers of Internet users. And, indeed, both countries' governments have found substantial ways to harness Internet use to their own ends. These include "defensive" measures such as forcing citizens and service providers to make use of proxy servers, blocking the access of these proxy servers to overseas (and domestic opposition) news sites, hiring censors to remove offensive material from bulletin boards, and closely monitoring Internet use at cyber cafés (Kalathi and Boas 2001; Rodan 1998). They also include "offensive" measures such as setting up government-sponsored portals, news sites, and discussion forums to try to mold public opinion in the government's direction (Kalathi and Boas 2001). For the most part, China and Singapore have neutralized the Internet as a tool for public opposition. Yet, at the same time, both countries are slowly becoming more open, with political restrictions less harsh than in the past. This is due not solely to the expansion of the Internet but to broader socioeconomic changes of which Internet expansion is a part.

I would contend that this gradual opening of political space is a natural process and speaks to an important relationship between democracy and the Internet. As pointed out by Kranzberg (1985), technology is not good, bad, or neutral. The Internet cannot automatically be assumed to have "good" effects such as democratization or "bad" effects such as aiding government control. Certainly, the Internet can be put to either of these uses. But this does not mean that the Internet is inherently good or inherently bad. Of course, to complicate matters, neither is the Internet neutral. Rather, it has certain affordances based on its history and design. One of the most important affordances of ICT is that it greatly increases the benefit-to-cost ratio of horizontal, networked communication. For institutions to fully exploit the Internet, they need to take advantage of this particular affordance. Of course, the Internet can also be used in other ways, such as narrowcasting material from the top to a passive audience below. But if these narrowcasting (or broadcasting) features of the Internet are exploited without also making use of the opportunities for many-to-many mass communication, the full advantages of Internet adaptation will not be gained. This applies at many institutional levels, from business to education to society.

Shoshana Zuboff (1988), for example, carried out a five-year ethnographic study of eight large businesses in the United States that were adopting information and communication technologies. Zuboff noted that initially employers expected computers to help them automate their tasks, but that while automation effectively hides many operations within the overall enterprise, information technology instead illuminates such operations. In other words, information technology improved productivity not so much by removing information and control from individuals (as in automation) but rather by expanding access to information and control by individuals. Zuboff used the word *informate* to describe this process. Zuboff's study showed that firms that were able to make the shift from automating to informating processes—by learning how to divest more authority and control throughout the workplace—were best able to take advantage of the information revolution, whether measured by increased productivity, smoother operations, or satisfied employees. And those firms that were not able to make the change faced problems. As a mill worker in Zuboff's study explained, "If you don't let people grow and develop and make more decisions, it's a waste of human life—a waste of human potential. If you don't use your knowledge and skill, it's a waste of life. Using the technology to its full potential means using the man [sic] to his full potential" (Zuboff 1988, 414).

Similar results have been found in educational research. As discussed in chapter 5, democratization of the classroom, school, or institution is not the only element leading to effective use of information technology, but it is an important element. Indeed, my earlier analysis of the Egyptian Ministry of Education can be interpreted largely in terms of the difficulties inherent in trying to spread ICT without bringing about democratization of an institution.

A similar process takes place at the level of societies and governments. To fully exploit the Internet for social and economic development, countries need not only to extend physical access to computers and connectivity but also to informate their societies. In other words, they need to expand power and control to individuals. This is in fact the dilemma faced by Singapore today, a country where widespread Internet availability and use have not yet had as broad an impact on the retooling of the economy as government leaders had hoped (indicated, for example, by the small ICT industry in Singapore and the limited Internet content production there; see Warschauer 2001c; Zook 2001c). An active debate is currently being waged among Singapore's government and economic elite regarding how to address this situation (see Yeo and Mahizhnan 1999), and in the meantime, Singaporean censorship of the Internet appears to be gradually decreasing.

In China the circumstances are somewhat different because the government is in a more volatile situation. At the same time, China has not yet faced the contradiction fully, because Internet diffusion, while accelerating, is still at a low per capita level. In the long run, if China is to continue its fast-paced economic development, the Internet will have to extend to new parts of the population, both sectorally (e.g., to the working class) and geographically (e.g., to the western, more impoverished parts of the country). At that point, the contradiction between the Chinese government's economic goals (requiring open information and interaction) and political approach (requiring media and communications control) may well come into more open conflict.

In summary, the Internet by itself does not bring about democratization or openness, but its diffusion does create new openings to struggle for democracy. How these opportunities will be realized depends to a large extent on popular action, so I now examine the impact of the Internet on voluntary associations and civil society.

Mesolevel Social Capital: The Power of Civil Society

Economists and social theorists point to a midlevel type of social capital between the microlevel of an individual's personal networks and the macrolevel of governmental institutions. There are various interpretations of mesolevel social capital, some of which include the role of corporate units (e.g., Turner 2000); however, for the purposes of this discussion, I focus on voluntary associations and civil society (see Woolcock 1998).

Civil society encompasses the networks, groups, organizations, and forms of association that exist between the private sphere and the state. It involves "*citizens acting collectively in a public sphere* to express their interests, passions, and ideas, exchange information, achieve mutual goals, make demands on the state, and hold state officials accountable" (Diamond 1994, 5, emphasis in original). According to Diamond, civil society performs a variety of vital democratic functions. First, it serves to monitor and restrain the exercise of state power, checking abuses and violations of the law and subjecting governments to public scrutiny. Second, it supplements the role of political parties in stimulating political participation on the part of the citizenry, increasing the political efficacy and skill of democratic citizens, and promoting an appreciation of the obligations and rights of citizenship. Third, civil society can be a crucial arena for the development of other democratic attributes, such as moderation, tolerance, a respect for opposing viewpoints, and the willingness to compromise. Fourth, it provides a vehicle beyond political parties for the articulation, aggregation, and representation of interests; this is particularly important for marginalized groups whose voices are not well represented by established political structures. Fifth, by allowing a range of cross-cutting issue-based movements to arise, civil society can help mitigate the polarities of political conflict. And finally, a democratic civil society can be critical for recruiting and training new political leaders beyond those who might emerge solely within political parties.

Alexis de Tocqueville, a nineteenth-century political theorist who wrote extensively on the role and value of citizens' associations, pointed to two technologies that are vital for their success: the meeting hall and the newspaper. The former provides citizens with an opportunity to directly communicate with each other, share opinions, form human bonds, and organize plans. The latter allows ideas to be broadly projected to association members and supporters. In his day, at least, it seemed true that "nothing but a newspaper can drop the same thought into a thousand minds" (Tocqueville 1835, 119).

However, both the meeting hall and the newspaper face important limitations today (see Klein 1999). A meeting hall can only bring together those people who either live in an immediate area or who can afford the time and expense to travel from a distance. Spatial and temporal barriers may make mass participation difficult, and the logistics of large meetings may make it difficult for all to have their voices heard. The costs of holding or attending meetings may also deter participation, especially among groups and individuals who are struggling financially. Finally, the many comments and ideas made at a meeting are often lost or subject to differing recollection, since they are not usually all recorded in an easily accessible form.

A newspaper overcomes some of these barriers only by substituting new ones. While a newspaper overcomes difficulties of space and time, it allows no forum for rapid interaction among group participants. Cost and publishing barriers also limit the extent to which it can serve as a medium of many-to-many communication.

In this light, the Internet can be seen as almost a miracle technology for citizens' groups. In a single low-cost technology it merges the roles of the meeting hall and the newspaper, and overcomes the limitations of each. "Meetings" can now be held by a limitless number of people all over the world on an ongoing basis, with comments and ideas automatically recorded for further analysis and discussion. Electronic newspapers, e-zines, and online discussion forums can drop a single thought into thousands of minds and also allow each of those thousands to interact immediately by replying to the author or to others.

That, at least, is the idea. How are citizens' groups actually using the Internet, and does this serve to empower those who are most marginalized? I examine two types of online communication: that among nonpolitical organizations and groups, where the primary benefit comes from exchange of information and social support, and that which is more explicitly political, where the goal is to aggregate and express demands.

Nonpolitical Association Online

The Internet potentially provides a valuable medium of communication for geographically dispersed people with shared interests. Everyone from sports fans to pet owners to alcoholics can go online to share information, find social support, and simply think together—a process referred to as *collective cognition*. Agre (1999a) explains,

Collective cognition in its various modes is greatly facilitated by the various community-building mechanisms of the Internet. Ideologies can form in the networked community of computer programmers; news can spread in the networked community of nurses; experiences can be shared in the networked community of cancer patients; patterns can be noticed by the networked community of pilots; agendas can be compared by the networked community of environmental activities; ideas can be exchanged in the networked community of entrepreneurs; stories can be told within the networked community of parents; and so on.

As Agre also points out, the opportunities for online networking provide special opportunities for marginalized groups, allowing them to level the playing field in situations that would normally disadvantage them. Doctors, business owners, and homeowners may already have plenty of channels of communication to protect their relative privileges, whereas patients, workers, and renters might need the extra affordances provided by the Internet. In some situations, the Internet's most important role may be to allow people simply to find each other. For example, large numbers of gays in China are said to have made contact with one another through U.S.-based Internet sites, and this in turn has contributed to a broader social assertiveness among, and public acceptance for, Chinese homosexuals (Pomfret 2000).

As an example of associational networking among groups in society, I briefly discuss female patients' Internet use in the United States. The United States is alone among wealthy industrialized countries in its lack of a national health care system. Approximately one in five Americans lack any health insurance, and those who are insured face a complicated network of opaque regulations, permission procedures, reporting regulations, and payment schedules that makes accessing health care a serious challenge. A growing number of Americans are insured via health maintenance organizations (HMOs) that hold down their costs through systems of complex barriers that restrict patients from obtaining expensive medical treatment. Indeed, a lack of knowledge about how the health care system works can in many cases be a matter of life and death because patients are too often discouraged (either directly or by bureaucratic roadblocks) from obtaining the health care they need.

This system is characterized by a vast gap between doctors and patients. Medical doctors in the United States are highly paid but are usually under various sorts of managerial pressure to limit their contact with patients or the information they provide them. HMOs often put doctors on tight schedules, rationing the time they can spend with individual patients in order to cut back on expenses. Doctors are also discouraged from providing too much information, either because this may take up too much time or because it might encourage the patient to seek expensive-to-the-insurer alternative tests or treatments. Doctors' fear of malpractice lawsuits also can discourage them from providing direct and clear information that is not hedged by legalese language.

Women in the United States are particularly burdened by this hard-to-negotiate health care system. On the one hand, women must take responsibility not only for their own health but also, in most cases, for the health of their children (and often other family members). Doctors in the United States, on the other hand, are disproportionately male. The result is a frequent gender gap in doctor-patient communication (Fairclough 1989; Tannen 1994; West 1990). In addition, women generally have lower incomes than men and frequently participate in part-time work. This means that women are more often without health insurance or have greater financial difficulties in covering medical expenses. In these circumstances, it is not surprising that women have turned to the Internet in large numbers to seek information, support, and solidarity about their health concerns.

A number of studies have found that health sites are among the most widely used on the Internet. A 1998 survey found that 46% of online users sought information via the Internet about a medical or personal problem (Green and Himelstein 1998). Health chat rooms and discussion groups are also widely visited (Lamberg 1997). Within the Yahoo! Groups discussion forums alone, there are more than 25,000 e-mail lists devoted to topics of health and wellness.[18] In addition, there are probably an equal or greater number of health-related Web sites, many of which combine online forums with links to background articles, testimonials, and other information.

Online forums can play a critical role in people's health. Studies have indicated that the amount of emotional support a patient receives is associated with fewer declines in social function and fewer medical problems (Bloom 1982; Dimond 1979). Patients can turn to the Internet when other sources of support are not readily available. Turner, Grube, and Meyers (2001, 235) explain this process:

As an illness constitutes an uncontrollable event that may influence several domains of an individual's life (income, contact with others, sense of achievement, physical capacity), support from others that responds to each of these domains can help achieve optimal adjustment.... Unfortunately, available support from others who understand the impact of the illness on these various life domains is not always available. Individuals may not be able to attend a support group targeted at their specific illness. However, with the advent of online support communities addressing specific concerns within individuals' lives, and the thousands of participants within these communities, the mathematical probability of a person finding someone with the same illness and treatment alternatives increases exponentially. For example, a patient can return home from being diagnosed, log in to an online community concerning the diagnoses, and ask about treatment alternatives or just express concern. Within minutes, that patient can receive specific responses to the posting. Similarly, the patient can learn about the diagnosis simply by reading the discussions taking place. The participants within online communities provide receptivity, interest, and disclosure, despite that they are strangers otherwise, because they can share a critical commonality. Therefore these large groups provide a strong probability that participants can find one or many other individuals who share similar specific symptoms treatments, reactions, problems, and challenges.

The value of these online support communities ties in with the theory of weak social ties discussed earlier in the chapter. Weak-tie relationships exist outside the dynamics or pressures of family relationships and are often contextual in nature (Adelman, Parks, and Albrecht 1987). The relative anonymity and objectivity provided by weak-tie relationships can thus provide an important alternative source of social support beyond that offered by family members.

Turner, Grube, and Meyers (2001) carried out a study of online communication on seven cancer-related e-mail discussion lists. Their study found that a 71% of the participants on the lists were women and that the amount of time people spent on the lists was inversely proportional to the amount of support they received from face-to-face partners. During the occasions when participants reported a lack in face-to-face support, they spent a greater amount of time online.

I spent several months investigating and participating in infertility online discussion forums during 2001. There are hundreds of such forums available on the Internet,[19] and they are overwhelmingly female in composition, even though infertility is a problem that affects both men and women. I found in these forums a highly ritualized environment, complete with its own elaborate set of code words, abbreviations, specialized greetings, and rules and regulations to guarantee a supportive environment. The nature of the discussion on these forums indicated the confluence of gender, class, and social issues that confront women with infertility problems. Common themes include exchange of medical information (e.g., trying to decipher what ought to have been explained clearly by doctors); investigation of medical clinics (e.g., trying to identify those whose staff are supportive, reliable, and competent); discussion of financial options (e.g., formulating strategies for financing expensive procedures that are rarely covered by insurance in the United States); and the sharing of all sorts of social support (badly needed in a culture in which women are often stigmatized for an inability to have children). One of the most common rituals is the formation of "cycle buddies," that is groups of women who are going through the same fertility procedure at roughly the same time who share notes and offer solidarity and support. These types of groups would be virtually impossible to organize on a face-to-face basis, and provide an example of how the Internet can facilitate access to new forms of social capital.

Involvement in online forums and networks does not replace the support of family and friends. Usually, online involvement focuses on a particular topic or shared interest rather than offering the more general type of support that comes from strong social ties. The person who offers a virtual hug on a breast cancer listserv is not likely to drop over to babysit, give you a ride to the auto mechanic, or invite you to the movies. Yet online involvement does provide an opportunity for creating and maintaining all sorts of beneficial horizontal associations that make our lives and our society richer and provide valued resources for those most in need.

Political Association Online

Though all communication takes place in a political context—and what could be more political than the U.S. health care system?—many forms of online association, such as those just described, do not have explicit political agendas. I now turn to the use of the Internet by nongovernmental organizations (NGOs) and popular movements with more explicit political or social agendas. The question driving this section is, Does the Internet help level the political playing field by extending greater communications and organizational power to those who lack it?

At first glance, the answer would seem to be yes. The past two decades have witnessed a virtual explosion of third-sector (nonstate, nonbusiness) political and social activity. According to one scholar of this movement, the long-term impact of this global "associational revolution" may prove to be "as significant to the latter twentieth century as the rise of the nation-state was to the latter nineteenth century" (Salamon 1994, 109).

And ICT has played an important role in the development of this third sector. International NGOs have used the Internet to share documents and coordinate strategies and campaigns. Grassroots groups use the Internet to mobilize their members and organize protests. And potential new activists are reached with information and articles on Web sites.

Probably the largest and most significant international social movement that has benefited from the use of the Internet is the antiglobalization movement. International use of the Internet by antiglobalization groups dates back to the 1980s, when international NGOs such as Greenpeace developed global communication networks for their staff. In 1990 the Association for Progressive Communications was born as a global network of social activists, created several years before the development of the World Wide Web.

The first major Internet-based campaign of the antiglobalization movement came in 1994, when the Zapatista rebels of Southern Mexico launched an armed rebellion. As pointed out by Ronfeldt et al. (1998), the Zapatista movement comprises three layers: (1) a social base of indigenous Mayan groups with a wide array of grievances against the Mexican government; (2) the Zapatista leadership of middle-class intellectuals who went to Chiapas in southern Mexico to create a guerrilla army and who articulated an antiglobalization agenda (e.g., militant opposition to the North American Free Trade Agreement and U.S. investment in Mexico); and (3) Mexican and international NGOs who regularly mobilize in support of Zapatista aims and campaigns.

From the beginning, the Zapatista leaders showed great skill in exploiting the Internet and other international media for their cause. Although popular images of the Zapatista leader, Sub-Comandante Marcos, uploading communiqués from his laptop deep in the jungle are almost certainly an exaggeration, the Zapatistas did coordinate with their supporters in Mexico City to have their materials typed or scanned into electronic format and distributed via the Internet. Within a short time after the rebellion had started, an unofficial Zapatista Web page was set up on the Swarthmore College Web server in the United States and a large number of Zapatista e-mail lists and Web sites were established throughout Mexico (Cleaver 1998). In 1995 the Zapatistas organized a national and international plebiscite to seek feedback from their supporters. Some 81,000 people from outside Mexico reportedly participated in the conference, mostly via the Internet (Cleaver 1998). E-mail communication later became key to the logistical and political planning of two intercontinental meetings, one of 3,000 people in Chiapas, Mexico, in 1996 and a second, of 4,000 people, in Spain in 1997, which together assembled grassroots activists from some forty countries.

The Internet proved critical for coordinating support among a diverse array of groups. The NGO network supporting the Zapatistas included issue-oriented NGOs, such as those supporting indigenous rights (e.g., the International Indigenous Treaty Council); human rights (e.g., the Inter-Church Committee on Human Rights in Latin America); and sustainable development (e.g., Food First). Especially important were infrastructure-building and networking-facilitating NGOs such as the Association for Progressive Communications and the U.S.-based PeaceNet (Ronfeldt et al. 1998).

Despite such widespread support, the Zapatista movement eventually lost steam, as neither its antitrade ideology nor its violent tactics could sustain substantial long-term support from its immediate social base or from the broader Mexican left. Nevertheless, in its few years of peak activity, the Zapatista uprising—by spawning an internationally coordinated online solidarity network—helped give birth to a much stronger antiglobalization movement, now linked and coordinated through the power of the Internet.

The next major Net battle of the antiglobalization movement arose in the period 1997–1998, following the World Trade Organization's proposal of an international trade accord called the Multilateral Agreement on Investment (MAI). In 1998 a draft proposal of the MAI was leaked onto the Internet, where it spread like wildfire. Although the MAI was largely overlooked by the mainstream media (leading a group called Project Censored to award the MAI its annual prize for corporate media "self-censorship"), an alternative Internet-based media campaign picked up the slack (Redden 2001). At least twenty anti-MAI coalitions launched Web sites opposing the accord; scores of other Web sites were initiated by advocacy groups, trade unions, and individuals (Smith and Smythe 1999). Local protests were organized in many countries, and in the end the MAI was never adopted. Though the demise of the MAI may have had more to do with the objections of potential signatories (e.g., France) than with grassroots campaigns, the movement did serve to further mature the Internet-enabled antiglobalization movement and whet its appetite for future battles.

Those were soon to come. Meetings of the World Trade Organization (Seattle 1999), the World Economic Forum (Melbourne 2000) and the World Bank/International Monetary Fund (Prague 2000, and Quebec 2001) provided new opportunities for antiglobalization groups to mobilize their forces. And once again, the Internet played a critical role. The Internet proved a perfect medium for organizing demonstrations that— unlike the large peace and civil rights protests of the 1960—had neither formal leadership nor unified demands. Rather, these protests became huge networked umbrellas to bring together a diverse array of organizations and individuals from Teamsters to anarchists, each with their own gripes and viewpoint. Umbrella Web sites, some of them anonymous, became critical in coordinating these disparate groups and fulfilled the dual role of Tocqueville's meeting hall and newspaper. As Redden (2001) explains, discussing a Web site used for the Melbourne protest:

The anonymous website S11.org provides a good example of how an Internet presence can increase grassroots communication and enhance the ability of protestors to mobilise. Along with related email discussion lists, it was instrumental in organising the large demonstration at the meeting of the World Economic Forum in Melbourne between September 11 and 13, 2000. Affinity groups from around Australia were able to co-ordinate their plans through the information provided by the Melbourne S11 Alliance online. Not only was information available at all times for groups in all locations, but its hypertextual form allowed easy, continual editing to reflect developments, while an overall structure with which users became familiar was maintained. The multimedia aspects of the site enhanced the production values to a level beyond that normally associated with activist literature. The first level sections were "home", "events", "call–2-action", "Melbourne groups", "regional groups", "organising", "accommodation", "protesting tips", "propaganda", "what is the WEF", "corporate profiles", "issues", "FAQs" and "links". These sections provided not only organisational information for those already committed, but also persuasive critiques of corporate globalisation, supplied by affiliates. During its time of peak operations it had one of the highest hit rates for any Australian website.

Eventually, permanent Web sites emerged to inform and coordinate the antiglobalization movement between its international demonstrations.

The best known and perhaps the largest is that of the Independent Media Center.[20] With background pieces, discussion forums, an online newspaper, links to listservs, a searchable archive, news updates provided by e-mail, and a sophisticated collection of photos and video, the site serves as a one-stop organizing center for the antiglobalization movement. Visitors to the site in September 2001 could find immediate links and information about nine major antiglobalization protests organized for October and November of that year.

For the antiglobalization movement and other resistance groups, the Internet does not supersede other forms of communication but supplements and expands them (see De Vaney, Gance, and Ma 2000). Redden (2001) discusses the mutually reinforcing use of new and old media by the antiglobalization movement:

The radical information published on the Internet is often available in print form, and activists typically use the Net as one medium among others through which they may work. Online activism often ties in with print media, with journalists sourcing information from websites, and magazines and leaflets quoting URLs. Many of the large convergences of people in demonstrations against corporate globalisation that have taken place since Seattle owe their scale to online organising between geographically dispersed interest groups. In these cases the Internet is used as a kind of metaconnection between more traditional local-level organisational activities such as meetings, telephone trees, leafleting, and posting flyers and stickers. Not only does online synergy translate into bodies on very "real" streets, but the protests in turn have been instrumental in bringing the critique of globalisation onto the agenda in broadcast media. So in exactly what medium does "reality" reside? Online networking has brought new possibilities of fast translocal mobilisation to the culture of activism, rather than having done away with all of its old elements. It seems to me to be more than a coincidence that the evolution of the Net has been continuous with the formation of new activist alliances on the basis of already existing activist communities. The Net has enhanced relationships between geographically dispersed and issue-based groups. As Roland Bleiker argues, "The phenomenon of speed has not annihilated dissent". It has transformed it.

All this would seem to suggest that the Internet is an unadulterated positive for the development of civil society and for the defense of marginalized groups. Certainly, the antiglobalization movement would not have achieved its current impact without the Internet. However, there are drawbacks to consider. The first, and most important, is that the Internet as a political tool tends to privilege those who have most access to it. This is not that critical a matter for nonpolitical associations, which are to a large extent non-zero-sum. For example, if cancer patients benefit from increased social support, that benefits society as a whole and harms no one. However, political activity can be considered zero-sum. Anything that strengthens one group's ability to promote its agenda thereby weakens the interests of those social forces with an opposing agenda. Looking again at the antiglobalization movement, the Zapatistas never used the Internet with their own indigenous base but mostly with its white, middle-class supporters in North America and Europe. This same relatively privileged social group—with its extensive access to computers and its high degrees of education, literacy, and English-language skills—has continued to dominate the antiglobalization movement. It is not an accident that antiglobalization protests focus to a large extent on protectionist issues (e.g., opposition to genetically modified agricultural products) that reflect the concerns of the protesters' own social class and cultural backgrounds. Other globalization-related policies that might benefit the poor in developing countries, such as ending rich-country tariffs on developing nations' imports, are given short shrift in antiglobalization protests. In other words, as long as unequal access to ICT persists, the expansion of the Internet as a political tool is as much a threat to the poor as it is an opportunity. The same principle applies here as in online voting: a disproportionate expansion of the communicative power of one demographic group can harm the interests of those groups whose voices have not been equally amplified.[21]

In addition, and as discussed earlier, the Internet favors weak social ties rather than the stronger ties that come from face-to-face communication. These weak ties are ideal for the types of "segmented, polycentric, ideologically integrated networks (SPINs)" that have emerged recently (Levine 2001, para. 22). These include the Zapatista movement, the antiglobalization movement, or even the international network of neo-Nazis, But, as Levine points out, these SPINs are not necessarily as solution-oriented as the kind of face-to-face movements that dominated earlier eras, such as the civil rights movement. As he explains,

SPINs need neither money nor enforceable rules; instead, technology minimizes transaction costs and shared values motivate members. SPINs have indeed protested and disrupted other institutions, but I doubt that they can devise (let alone implement) positive programs. For instance, the anti-globalization move-ment has put protesters on the street, but it seems incapable of creating a new system of international trade. (Para. 22)

The Internet also allows for an extreme form of narrowcasting and information filtering that might go against the interests of civil society. In reading a newspaper or watching the television news, citizens will come across issues and ideas that might surprise them. In contrast, the Internet allows people to restrict their news and communication chan-nels to extremely narrow fields, and actually can discourage the kind of tolerance and informed compromise that comes from exposure to a wide range of ideas and people (Sunstein 2001).

Finally, even in situations when the Internet does provide political space for persecuted groups, it may also create new tools for surveillance and infiltration. For example, at one stage the Internet provided a pow-erful means for the Falun Gong to clandestinely organize in China (O'Leary 2000), but it later provided an equally powerful means for the government to track down and arrest Falun Gong members (Rosenthal 2001).

Conclusion

Whether at the micro-, macro-, or meso-level, preexisting social capital can have an important influence on individuals' and groups' ability to use ICT, and if properly exploited, ICT can be promoted in ways that encourage the development of social capital. Strategies that take into account the social nature of access, recognize the interaction between face to-face and online communications, and combine Internet use with a broad range of other new and old media provide the best opportuni-ties for promoting social inclusion through use of ICT.

7

Conclusion: The Social Embeddedness of Technology

Two of the most astute analysts of the sociology of the Internet are Paul DiMaggio and Eszter Hargittai of Princeton University (see, for example, DiMaggio et al. 2001; Hargittai 1999; 2002; in press). In a recent piece, DiMaggio and Hargittai (2001) discuss the issue of the digital divide. They assert that now that Internet diffusion rates have increased, scholars should shift their attention from the digital divide—inequality between haves and have-nots based on dichotomous measures of access—to digital inequality, by which they mean differences among people with physical access to the Internet. Digital inequality, from their perspective, encompasses five main variables: technical means (inequality of bandwidth); autonomy (whether users log on from home or at work, monitored or unmonitored, during limited times or at will); skill (knowledge of how to search for or download information); social support (access to advice from more experienced users); and purpose (whether they use the Internet for increase of economic productivity, improvement of social capital, or consumption and entertainment).

I agree with DiMaggio and Hargittai's points but believe they need to be extended. Specifically, the broad, multifaceted approach they bring to bear on issues of digital access and equality needs to be applied not only in situations in which Internet penetration is high but also in situations in which it is low and just beginning. Indeed, it is precisely in such situations that the promotion of skills, social support, and autonomy, while carefully paying attention to the underlying purpose, can be most important.

An example will illustrate the point. During a research trip to Latin America, I met with the president of a national telecenters project. The president, a Westerner, had traveled to this particular Latin American country to establish the project on behalf of the foundation for which he worked. He and the foundation were determined to set up a large number of telecenters as quickly as possible so as to bring computers and Internet access and training to great numbers of people. By demonstrating results, they hoped to impress the business and philanthropic communities in their home country and thus raise further funds for the project.

The president explained to me that in order to achieve the goal of setting up many centers it was necessary to partner with local community organizations, which entailed a process of contact, communication, discussion, and negotiation. He made clear how distasteful he found this task. As he explained, he simply wanted to get his centers up and running as soon as possible, and he had little patience with the slow and tedious process of building local partnerships and addressing people's questions and concerns. He also explained to me his pride that his project was nonpolitical and told me that other telecenter projects in the country were hindered by their political agendas. Finally, I could see that he kept a tight rein on the overall project; though local staff was involved in helping run the project, he—a newcomer to the country without prior experience in the region—maintained the role as the project president and carried out much of the local negotiations.

I was able to visit one of the project's telecenters—in fact, the only telecenter it had managed to set up thus far. The telecenter had been established at the office of a local community organization in a poor neighborhood of a large city. There were two computer labs there, one for training on computer skills and one for Internet access. The "nonpolitical" nature of the training was evident. In contrast to the Schools for Information Technology and Citizens' Rights (see chapter 5), this telecenter appeared to devote no attention to integrating social or community issues into the curriculum. The training I witnessed consisted of decontextualized exercises. The course offerings were all based on particular software programs rather than on use of these programs for meaningful ends. There was not much evidence of involvement of the community organization in helping define the instructional curriculum or the direction of the telecenter project, although some of the community organization's staff members were deployed to hand out elaborate brochures the national project had developed. At least judging by that evening, community response was poor; only a few people showed up for computer training, and I didn't witness any show up for Internet access.

I spoke to the coordinator of the partner community organization, a woman who had patiently built the neighborhood organization over two decades. She expressed strong concern about the direction the telecenter was taking. Referring to the Western project president, she told me, "he wants to reach numbers, we want to reach people." When I asked her what she meant, she explained that he was aggressively pushing them to deliver any kind of training to as many people as possible as quickly as possible in order to show results to outside funders (a point that the president himself confirmed), while she and her organization were more interested in taking the time to design and implement programs that would really make a difference in people's lives.

In all fairness, it is important to point out that real-world contradictions do arise between reaching "numbers" and reaching "people," that is, between reaching larger numbers of people in quicker and less intensive ways and reaching smaller groups of people in slower and more extensive ways. These are trade-offs that any type of social development project must weigh, and the need to deliver results to funding sources is one consideration that sometimes has to be taken into account. However, in this case, no such trade-off was witnessed; by emphasizing numbers *instead of* people's real needs, the project failed to reach either people *or* numbers. The problem appeared to lie in large measure with the Western foundation trying to carry out its work in a "nonpolitical" fashion—in other words, in its *own* political fashion, which deemphasized the value of community participation and mobilization—without taking into account or addressing local political concerns or views.

As this example and others presented in this book demonstrate, social context, social purpose, and social organization are critical in efforts to provide meaningful information and communication technology (ICT) access, whether in developed or developing countries. Issues of what constitutes skill and how it is developed, what purposes are served by gaining access, who develops autonomy and how, and what kinds of social resources are mobilized are all crucial for promoting social inclusion. Since I have covered wide ground in making these points, I now try to tie the diverse threads together by reviewing the question that has driven this book, this time from a theoretical perspective.

Social Embeddedness of Technology

The framework of the digital divide implies that technological and social contexts can be separated from each other and that these two separate contexts interact through a mechanism of causality. Programs are thus designed to solve the technological problem is the belief that this will ameliorate one or more social problems. This separation is seen conceptually in one of two ways: determinism or neutralism. Media determinism characterizes technology as existing apart from society and exerting an independent impact on it. We are concerned, for example, about the impact of television on children, about the impact of computers on learning, and about the impact of the automobile on society. And, in each of these cases, we *should* be concerned about the role of technology, but none of these impacts can be analyzed outside of the particular social contexts at hand.[1] There is a complex mutually evolving relationship between a technology and broader social structures, and the relationship cannot be reduced to a matter of the technology's existing on the outside and exerting an independent force.

Those involved in educational applications of technology confront attitudes of technological determinism regularly. How many times have administrators or funders demanded to know the impact of computers on learning, without any consideration of the context in which computers are used or the purposes they are used for? This focus on the omnipotent machine, removed from considerations of use or context, has been criticized as the "fire" model of education technology, based on its notion that a computer in a classroom will automatically generate learning in the same way that a fire automatically generates warmth (Dede 1995; 1997).

Seemingly in opposition to determinism, but actually overlapping with it, are neutralist (or instrumental) theories of technology. From this perspective, technology is devoid of any particular content or values. Rather, it is a neutral tool, indifferent to the ends for which it can be employed (see Feenberg 1991). From this perspective, the computer is not particularly good or bad; it is just a piece of metal that can be used for any purpose.

This last position has a lot of commonsense appeal, and it is in part a corrective to determinism. But, like determinism, it fails to account for the social embeddedness of technology. Technologies may not be good or bad in themselves, but neither are they neutral (Kranzberg 1985); rather, they carry with them certain values based on their own history and design (Feenberg 1991).

The personal computer and the Internet, for example, emerged in a particular U.S. social context, and their designs reflect the values and perspectives of the American engineers who worked on them. For example, for character encoding, computer engineers developed and adopted the American Standard Code for Information Interchange (ASCII 2001; Jennings 2001). As a seven-bit system, ASCII only allows for the representation of 128 (2^7) characters; once lower-case and upper-case letters, numbers, and standard punctuation marks were included, there was no room left over for letters with diacritical marks (as are common in the alphabets of many European languages), or for characters and symbols of the diverse non-Roman alphabets around the world. Eventually a more flexible system called Unicode has emerged, but it has not yet become standardized in the same way that ASCII has. The limitations of ASCII help explain why English and other Romanized languages got a head start on the Internet, a bias that strongly influenced who has been able to access the Internet, what materials are published there, and what broader social systems and structures are privileged (see chapter 4).

Another bias of personal computing is the desktop interface, which is based on an office metaphor (e.g., files and folders) rather than on other possible metaphors (a kitchen, a tool shed, a farm), thus being more accessible to people with certain kinds of prior experiences (Burbules and Callister 2000; Selfe and Selfe 1994).

These are just two of the many ways that coding decisions—based on the social history of ICT and the viewpoints of U.S. engineers—have influenced the diffusion and accessibility of computers and the Internet. Many similar examples, including the relation of computer coding to issues of intellectual property, privacy, free speech, and sovereignty, are discussed by Lessig (1999). And there are far broader design issues than coding. For example, the social milieu in which ICT arose has also shaped a broad range of research and development decisions, affecting issues such as the cost of hardware and software, the difficulty of using computers, the range of devices that can be used for computing and online communication, and the transmission media of telecommunications. None of these can be considered neutral.

The significance of coding and other design issues once again demonstrates the complex interrelationship of technology and society in contrast to a simplistic notion of outside impacts. Whereas the deterministic view sees impact as inherent in the technology and the instrumental view sees impact as within the domain of the individual user, neither perspective captures the ecological intertwining of technology and society.

The limitations of media determinism and neutralism are illustrated by a discussion of the printing press. There is no doubt that European society changed a great deal in the centuries following the Gutenberg revolution. The affordability and wide diffusion of printed texts laid the basis for modern scholarship and science by making published data more readily available in thousands of copies. The publication boom of the fifteenth to seventeenth centuries aided the Protestant reformation by providing an alternative source of communication and authority to that of the Catholic hierarchy. Education was transformed, as teachers and students were relieved of the burden of slavish copying. Students who took full advantage of technical texts, which served as silent instructors, "were less likely to defer to traditional authority and more receptive to innovating trends" (Eisenstein 1979, 689). The very format of the printed book—with tables, figures, footnotes, and indexes—contributed to new ways of categorizing and conceptualizing information (Eisenstein 1979; McLuhan 1962).

However, while the invention of movable type made these changes possible, it did not autonomously bring them about. In fact, movable type was invented in China more than 400 years before its development in Europe (Carter 1925) but was little used in Asia. Its widespread diffusion in Europe depended directly on other changes already underway there, including the emergence of a capitalist class, colonialism, and "a heightened sense of individuality and personality, of nationalism and secularism" (Murray 1995, 28). Nor can the printing press be seen to have caused the spread of mass literacy in Europe because that did not occur until several centuries later. It was the industrial revolution, not the industry of printing, that brought about mass print literacy and helped shape its current characteristics (see Tuman 1992). In other words, rather than the printing press being introduced from the outside and having an impact on society, it emerged from the inside and interacted with other elements of society in an ecological fashion. As Neil Postman (1993, 18) put it, "Fifty years after the printing press was invented, we did not have old Europe plus the printing press. We had a different Europe".

This illustrates well the difference between what Levinson (1997) refers to as "hard" and "soft" media determinism. Hard determinism holds that technological change automatically causes social change, an assertion that is easily disproved by the example of the differential impact of movable type in Asia and Europe. In contrast, soft determinism enables social change but does not in and of itself bring it about.[2]

As exemplified by the history of the printing press, any technology—especially a major new medium of communications—does not exist outside a social structure, exerting an independent impact on it. Nor is it a neutral tool to be deployed in a haphazard fashion. Rather, technological and social realms are highly intertwined and continuously co-constitute each other in a myriad of ways. This co-constitution occurs within organizations, institutions, and in society at large.

ICT and Organizations: Sociotechnical Networks

The research tradition that analyzes the social embeddedness of ICT in organizations is knows as social informatics. This approach emerged at the University of California, Irvine, in the 1970s with a series of studies on the role of computerization in a wide array of organizations, including government agencies, factories, banks, schools, and offices (see Kling 1991). The research, conducted by teams of scholars from computer science, public administration, and political science, yielded important insights about the computerization of organizations.[3] Perhaps the most significant was that computing could not be understood as a separate tool but was rather part of an overall package. Kling and Scacchi (1979; 1982) explain the difference between a tool and a package model as follows:

> [The tool metaphor suggests that] one can safely focus on the device to understand its use and operation. In contrast, the package metaphor describes a technology which is more than a physical device. . . . The package includes not only hardware and software, but also a diverse set of skills, organizational units to supply and maintain computer-based services and data, and sets of beliefs about what computing is good for and how it may be used efficaciously. Many of the difficulties users face in exploiting computer-based systems lie in the way in which the technology is embedded in a complex set of social relationships. (1982, 6)

One of the key studies that informed this research agenda was a national investigation of computer use by local governments in 500 U.S. cities (Danziger et al. 1982). The study found that the equipment and facilities available were only a minor part of the impact of computer technology on local government. Much more significant was the organizational systems set up to regulate and control computing, and the visions, competing interests, funding mechanisms, and struggles of key actors, including managers, policymakers, vendors, employees, and citizens. The researchers concluded that the key to understanding the computing system of an organization was not what kinds of equipment and facilities it has but rather the kinds of things people do with them (9).

Social informatics later grew into a national and international research tradition (see Kling 1999; 2000). Scholars in this field investigate the ways in which technology-in-use and social worlds co-constitute themselves in highly intertwined fashion (Kling 2000). The original computing package model has broadened to the concept of sociotechnical networks, explained by sociotechnical models that Kling has nicely summarized (table 7.1).

The studies discussed in this book draw on and reinforce the concept of the sociotechnical network. They provide further evidence that looking at what people do rather than merely at what equipment they have is necessary to make effective use of ICT for social change and inclusion.

Table 7.1
Standard Models vs. Sociotechnical Models of IC

Standard (Tool) Models	Sociotechnical Models
ICT is a tool.	ICT is a sociotechnical network.
A business model is sufficient.	An ecological view is also needed.
One-shot ICT implementations are made.	ICT implementations are an ongoing social process.
Technological effects are direct and immediate.	Technological effects are indirect and involve different time scales.
Politics are bad or irrelevant.	Politics are central and even enabling.
Incentives to change are unproblematic.	Incentives may require restructuring (and may be in conflict).
Relationships are easily reformed.	Relationships are complex, negotiated, multivalent (including trust).
Social effects of ICT are big but isolated and benign.	Potentially enormous social repercussions from ICT (not just quality of work life but overall quality of life).
Contexts are simple (a few key terms or demographics).	Contexts are complex (e.g., matrices of businesses, services, people, technology, history, location).
Knowledge and expertise are easily made explicit.	Knowledge and expertise are inherently tacit/implicit.
ICT Infrastructures are fully supportive.	Additional skill and effort needed to make ICT work.

Source: Adapted from Kling (2000) with permission of *The Information Society*.

ICT and Institutions: Shaping the Structure of Relationships

The contributions of social informatics are greatly enhanced if we overlay them with an institutionalist perspective. The study of institutions has long been important to social theory, especially in the last thirty years, as a "new institutionalism" has swept through the fields of history, sociology, economics, and political science (DiMaggio and Powell 1991; Goodin 1996). Although this new institutionalism takes on different meanings in different disciplines, in the broadest sense it can be seen as an attempt to "blend both agency and structure in any plausibly comprehensive explanation of social outcomes" (Goodin 1996, 17). This in essence overcomes the false contradiction between determinism and neutralism, by focusing on how the influence of individual agents and of social structures are together mediated by institutionalized ways of thinking and acting.

An institution is different than an organization or a collection of organizations. Rather, it refers to the types of routinized interaction (also known as scripts and schema; see DiMaggio and Powell 1991) that typify and shape human activity in a defined realm. Institutions serve to structure relationships between people by inducing them to insert themselves into a particular order and way of interacting (Agre 2001a). For example, the institution of academia is not just a collection of universities but a way in which relationships between undergraduate students, graduate students, instructors, professors, and staff are formalized and structured. The tenure system, the scholarly conference, the job search process, and the dissertation process are just a few of the elements that contribute to the institution of academia.

Technologies do not exist apart from institutions, exerting an external impact, but are part and parcel of them. The institution shapes the workings of the technology while the technology shapes the workings of the institution. The microwave oven has become part of, and helped shape, the institution of "dinner" (at least in many countries), and the institution of dinner has in turn shaped the development of the microwave oven (e.g., by influencing the features that manufacturers have chosen to include). This techno-institutional interaction is especially powerful in relation to ICT because of its great generality and adaptability (Agre 1999b). Though a microwave oven heats food, ICT "is equally at home in offices, factories, trucks, telephones, shirt pockets, spacecraft, thermostats, intensive care units, and kindergartens" (para. 1). In addition, while all technologies serve to structure human relations (think, for example, how the microwave oven has facilitated the entry of women into the workplace), many do so as a by-product of their main function. In contrast, the very purpose of ICT is to restructure human communications and relations. For all these reasons, ICT is bringing about a "thorough renegotiation of the ground rules" of every institution (Agre 1999b, para. 1).

The notion of the institutional embeddedness of technology offers a better alternative than the concept of a digital divide or digital solution.

Returning to the theme of distance education (see chapter 5), the digital divide framework can lead one to oversimplify the potential positive contribution of distance education by taking at face value the notion that ICT can extend educational opportunities to previously excluded groups. While this is one possible outcome, it is not the only or necessarily most likely one. A more refined institutional analysis is required to evaluate the actual role that ICT is playing in academia. For example, one of the functions of academia is to sort out students, and an elaborate array of mechanisms (admissions, test scores, financial aid, advising, grading) and organizational forms (different tiers and levels of colleges and universities) are used to that end. Only when considering the institutional function of social sorting is it possible to evaluate the possible impact of distance education. As seen in the discussion in chapter 5, distance education is as likely to magnify this process of sorting as it is to undermine it, by amplifying trends toward more unequal higher education.

Similarly, efforts to make use of ICT to meet the needs of rural villagers should be based on analyses of relevant institutions, such as banking, health care, and local government. The starting point for a progressive consideration of ICT in any institution should not be the digital divide and how to overcome it but rather the broader social structures and functions of the institutions and how ICT might be used to help make them more democratic, equitable, and socially inclusive.

ICT and Society: Applying Critical Theory

Finally, a key related concept is a critical theory of technology (Feenberg 1991; 1999b; Winner 1986). Drawing on the broader critical theory of the Frankfurt school, Feenberg situates technology within the underlying unequal power relationships that exist in society. The bias of technology reflects these power relationships, as seen, for example, in how the Internet's historical bias for English reflects the social, political, economic, and technological power of the United States vis-à-vis other countries. The diffusion and use of technology are understood as a "scene of struggle" (Feenberg 1991, 14).

The importance of critical class analysis is seen, for example, by considering the role of ICTs in rural poverty alleviation in South Asia. As a recent essay points out ("ICTs" 2001), poverty alleviation in India is not a matter of service delivery, but one of enhancement of agency of the poor, based on the transformation of class, caste, ethnic and gender relations within which the poor exist. The "technology as solution" approach (Heeks 1999) ignores the social structures that determine both access and impacts. While cheapening technology certainly has a role to play in making it more acceptable to the poor, social structures are crucial in determining who is able to access any technology and use it beneficially.

The essay points out how social structures of poverty vary according to region in India. In the hill-forest areas, communities of indigenous people supply raw materials (e.g., timber) and ecosystem services (e.g., hydrological, biological). With almost the entire community suffering from poverty, class contradictions are minimal. ICT can be used to enhance people's skill and information so that they can improve their productivity (by learning more about the delivery of ecosystem services). In contrast, in the Indian plains, a major cause of poverty is landlessness, and huge contradictions exist between the landless poor and the large landholders. Projects that increase agricultural productivity—for example by providing information about market prices—may have a small trickle-down effect to the landless poor but cannot in themselves qualitatively undo the underlying problem of landlessness. For this to occur, information must be combined with mobilization, and ICT projects will be in the end most meaningful if they find ways to lend support to mobilization efforts (for example, by linking nongovernmental organizations that are active among the landless poor).

In summary, the organizational, institutional, and societal levels of analyses all overlap. Each points to the critical role of social structures in shaping how technology is diffused, and the corresponding importance of social analysis and goals in the planning of ICT development projects.

Technology for Social Inclusion

The concept of a digital divide has helped focus public attention on a critical social issue: the extent to which the diffusion of ICT fosters stratification and marginalization or development and equality. With the world's attention focused on this problem, it is now the time to put forward a more refined conceptual framework to the problem and a more informed policy and research agenda.

The overall policy challenge is not to overcome a digital divide but rather to expand access to and use of ICT for promoting social inclusion. The policy implications of this will vary according to circumstance, but I touch on a few issues here, summarizing some of the main themes discussed in this book.

First, analyses of the problem must begin with examination of social structures, social problems, social organization, and social relations rather than with an accounting of computer equipment and Internet lines. An accounting of equipment is part of the overall analysis, but a fairly small part; if interventions are designed to address social problems, they must be planned by focusing on the overall structures and relationships that give rise to those problems. Analyses must take into account not only social problems but also best social practices. Technology can often serve to amplify already existing practices; by examining how people in a particular realm currently learn, collaborate, share, and succeed, technological interventions can be sought that amplify these practices (Agre 1998).

Once social problems or goals are identified, programs should be based on a systemic approach that recognizes the primacy of social structure and promotes the capacity of individuals or organizations for ongoing social change through innovation of those structures using technology. Corea (2000) discusses this strategy in depth, pointing out that information technology implementations often create only superficial change, with organizations returning to their ingrained ways once the new systems have been "absorbed into the previous web of calcified inefficiencies" (9). Rather than just foisting technologies haphazardly on people, a better solution is to foster the "long-term nurturing of behaviors intrinsically motivated to engage with such technologies" with the goal of achieving "an 'innovating' rather than a 'borrowing' strategy of growth as a means to reduce technological disparities" (9). This can bring about a "catching up process" through development of capacity "in the generation and improvement of technologies, rather than in the simple use of them" (Perez and Soete 1988, p. 259, quoted in Corea 2000, 9). All of this requires changes in the social environment to facilitate "the learning of new behaviors that propagate continuous improvements in conditions of living" (Corea 2000, p. 9). This process of innovation might take many forms. Rural teachers might learn how to create their own technology-based materials based on local conditions rather than only using commercial software developed for other contexts. A crafts cooperative might learn how to develop and manage its own Web site rather than just posting its announcements on somebody else's. Nongovernmental organizations might learn to establish and run their own networks of telecenters rather than just attending cyber cafés.

In promoting such efforts and programs, it is essential to understand and exploit possible catalytic "effects" of ICT. Many important changes in social relations may come from the human interaction that surrounds the technological process rather than from the operation of computers or use of the Internet. For example, a new computer laboratory in a low-income neighborhood may also become a meeting hub for at-risk youth and college student mentors. Or the involvement of community members in planning the laboratory may bring together new coalitions that can also work for other types of community improvement. The social importance of ICT in the information economy and society means that ICT initiatives often have powerful leveraging potential that can be used to support broader strategies for social inclusion.

The roles of leadership, vision, and local "champions" (McConnell 2000, 8) are crucial to the success of ICT projects for social inclusion. A common mistake made in ICT development projects is to make primary use of computer experts rather than of the best community leaders, educators, managers, and organizers. Those who are capable of managing complex social projects to foster innovative, creative, and social transformation will likely be able to learn to integrate technology into this task. On the other hand, those with technological skills, but lacking understanding of the complex human issues at hand or the leadership ability to address them, will usually prove less effective (see Agre 1998).

The process of organizing, designing, implementing, and evaluating ICT projects must itself be open to innovation and flexibility. Good big things come from good small things, and room for innovation, creativity, and local initiative is critical to give the space for good small things to emerge. The discussion of education projects in chapter 5 pointed to the need for flexible pilot programs as part of the development process. Scalability is of course an important aspect, and the potential for scaling up has to be part of the formative and summative evaluation of pilot programs. But lock-step, centrally organized, large-scale initiatives with no room for local experimentation and innovation do not meet the needs of a rapidly changing information economy or society.

Finally, market mechanisms can be effective for expanding access to computers and telecommunications but they are not sufficient. Governments need to consider in what situations restrictions on markets are hindering expanded access, and take steps to end such restrictions. Evidence suggests that these steps should include removing import tariffs on computer hardware and software and ending monopoly control of telecommunications (Hargittai 1999; Wallsten 2001). At the same time, for a variety of reasons—including the limited purchasing power of the poor—market mechanisms will clearly be insufficient for providing universal access. Funds for research and development on low-cost computer and Internet alternatives, incentives for extension of telecommunications services to rural areas, and the backing of research on the causes and consequences of restricted ICT access are all ways that governments can improve on the power of markets without undermining them. Initiatives that harm the expansive potential of markets, for example, by prematurely locking in proprietary infrastructures, should be avoided.

Expanded Research

A technology for social inclusion approach also requires an expanded research agenda. Some important steps in this direction have already been taken. For example, the U.S. National Telecommunications and Information Administration has refined its measures of physical access to computers and the Internet to include more gradations of access (whether or not people have broadband Internet access) and to target additional populations threatened with social exclusion (e.g., the disabled). Such graded measures of a variety of populations should be encouraged in a wide array of research studies.

At the same time, research should be expanded in other areas of resources related to the availability of content in specific languages, the language choices made by people online, the skills and "electronic literacies" that users have, and the relation of community and social support to ICT use (DiMaggio and Hargittai 2001; Hargittai, in press; Hoffman and Novak 2001).

Beyond issues of access, it is also critical to study patterns and types of usage. For example, in school settings most research has thus far gone into measuring how much equipment and infrastructure schools have; much less has gone into examining how computers and the Internet are actually used with different school populations. These emphases should be reversed, with the larger priority given to examining use.

And beyond use, there is also the question of outcome. As DiMaggio and Hargittai (2001, 17) suggest, studies can "investigate variations in rates of return to technology use for different subgroups within the population." These rates of return can refer to a wide range of issues related to learning, emotional satisfaction, social capital, participation, income, and other forms of social or economic benefit. For example, one recent outcome study found that disadvantaged minorities pay the same prices as white buyers when purchasing cars online in the United States in contrast to the 2.1% price difference (+$500 USD) they pay when purchasing cars in person (Morton, Zettelmeyer, and Silva-Risso 2001).

It is especially important to supplement individual-level research with "analysis of institutional factors that shape and modify over time the relationships between individual characteristics and individual outcomes" (DiMaggio and Hargittai 2001, 17). For example, in researching telecenters, correlations can be examined between the types of location, facilities, ownership, administration, and purpose on the one hand, and the individual characteristics and outcomes of users on the other hand.

These expanded research goals will require a corresponding expansion of methods and approaches. Methods required will include observational designs, analyses of user behavior, cross-national comparisons, international surveys, and political-economic research on regulatory issues (DiMaggio and Hargittai 2001; Hargittai, in press). Particularly important will be longitudinal ethnographic studies that can reveal the ways in which social structure, technological innovation, and human development are intertwined, as Zuboff's (1988) study showed at an earlier stage of ICT diffusion. These types of qualitative studies of technology in social context will not always yield precise answers, but as one leading statistician noted, "far better an approximate answer to the right question, which is often vague, than an exact answer to the wrong question, which can always be made precise" (Tukey 1962, 13).

This proposed research agenda is broad and ambitious but not impossible. It will require strong disciplinary research methods (e.g., in sociology, anthropology, economics) together with equally strong interdisciplinary content knowledge of the wide variety of scholarly fields that issues of ICT access and use encompass. By bringing together teams of scholars from different disciplines, backgrounds, and cultures—and encouraging individual scholars to cross disciplinary boundaries— research on ICT in society can help bring about a new scholarly paradigm that is more in line with the imperatives of a postindustrial society in which knowledge and disciplines cannot be tightly bound. If flexibility, creativity, and many-to-many multimodal interaction are hallmarks of the information era, they will also be the hallmarks of the scholarship that the era demands.

Conclusion

Walter Ong, a prominent scholar of orality and literacy, once wrote that "technologies are not mere exterior aids but also interior transformations of consciousness, and never more than when they affect the word" (1982, 82). The statement has been criticized for being deterministic, but I don't find that to be the case. It simply refers to the fact that technology and the mind cannot be separated, as illustrated by the intertwining of the blind man's sensory perception with his walking stick (see chapter 5).

Just as technology becomes part of the neural network of the mind, it also becomes part of the social network of humanity. And never has this been more the case than with information and communication technologies, which function not only as the electricity of the twenty-first century but also as the printing press, library, television, and telephone, not to mention school, social club, mall, debating society, and gambling den. The Internet is not so much a tool as a new social space that restructures social relations (Poster 1997).

As researchers of ICT and its social context, we may sometimes tally up computers and Internet accounts; however, this is not an end in itself but rather part of a broader effort to better understand the process of technology use and the role of ICT in human and social development. Similarly, as social advocates, we may work to distribute computer equipment, but again as one step toward a larger purpose of helping people participate fully in the information economy and network society. That participation requires not only physical access to computers and connectivity, but also access to the requisite skills and knowledge, content and language, and community and social support to be able to use ICT for meaningful ends. The tasks are large, but so is the challenge: reducing marginalization, poverty, and inequality and enhancing economic and social inclusion for all.

Notes

Introduction

1. Information on this project comes from a paper by Sugata Mitra (1999); personal communication, July 2001, Chetan Sharma; and my own visit to the site and interviews with users and community residents in July 2001.

2. Personal communication, July 2001, S. Regunathan, Principal Secretary for Information Technology, Government of New Delhi.

3. Information on this competition and its results comes from <http://www.eircom.ie>; the Web sites of the four winning towns, <http://www.ennis.ie>, <http://www.castlebar.ie>, <http://www.kilkenny.ie>, and <http://www.killarney.ie/>; and personal communication, May 2001, John Mooney, University College Dublin.

4. Personal communication, May 2001, John Mooney.

5. Personal communication, January 2000.

6. For general overviews, see Askonas and Stewart (2000), Byrne (1999), and Littlewood, et al. (1999). For particular discussion of relationship to technology see European Commission (2001b).

7. Digital Divide discussion list, <http://www.digitaldividenetwork.org/>; Global Knowledge for Development discussion list, <http://www.globalknowledge.org/discussion.html>; Association for Internet Researchers discussion list, <http://www.aoir.org/mailman/listinfo/air-l>; and Red Rocker Eater News, <http://dlis.gseis.ucla.edu/people/pagr/rre.html>.

Chapter 1

1. According to the *New York Times* (Labaton 2001), Michael K. Powell, the chairman of the Federal Communications Commission, "said he thought 'digital divide' was a dangerous phrase because it could be used to justify governmental entitlement programs that guaranteed poor people cheaper access to new technology. 'I think there is a Mercedes divide,' he said. 'I'd like to have one; I can't afford one.'"

2. The European Union, in particular, has embraced the notion of technology for social inclusion; see, for example, European Commission (2001b). The federal government of Brazil has launched a national digital inclusion project; see Governo Electronico (2001). In the United States, the federal government's National Telecommunications and Information Administration (NTIA 2000), which first popularized the term *digital divide*, has also shifted to the terminology of digital inclusion.

3. I am greatly indebted to the groundbreaking work of Manuel Castells (1993; 1997; 2000a; 2000b; 2001) on the informational economy and the network society, and I draw extensively from his research and sources.

4. Revenue rankings from <http://www.fortune.com> (December 12, 2001). For examples of reports on Dell's business model, see Magretta (1998), Wysocki (1999), and DiCarlo (1997). For the company founder's view, see Dell and Fredman (1999).

5. For details on the index, see Theil (1967).

6. Income figures based on Purchasing Power Parity (PPP), a measure that examines how much can actually be purchased with local currency. In actual U.S. dollars, the income figures would be much lower.

7. These programmers were known as hackers, originally meaning someone who had the skill and patience to code enormous programs (Spyd3r 1998).

8. Electronic communication of research findings is especially prominent in physics, mathematics, and computer science. The "e-print arXiv" (<http://arXiv.org/>), funded by the U.S. National Science Foundation and consisting of papers posted by researchers in these fields, currently includes 170,000 papers and is adding about 35,000 new papers per year (Ginsparg 2001). In some other fields, attempts to set up electronic archives of research have met resistance. For an in-depth discussion of a highly contested effort to set up an electronic archive of biomedical research, see Kling, Fortuna, and King (2001).

9. S. Amadeu da Silva (2001, 30). I have provided the translation; italics in the original.

Chapter 2

1. I use the term "Hispanic" throughout the book when citing or discussing studies that use that term. Otherwise, I use the term "Latino."

2. The approach to literacy that emphasizes its social context is known by many names, including new literacy studies, social literacies, sociocultural approaches to literacy, ideological approach to literacy, and critical literacy; see discussion of these terms in Gee (1996), Lankshear (1994), and Street (1993).

3. Exceptions abound, of course, especially when one takes a broad view of literacy that considers the ability to make use of texts to understand and take action

in the real world. See, for example, Mastin Prinsloo and Mignonne Breire's (1996) edited collection about literacy practices in South Africa.

4. Data are for 1999 and are taken from the United Nations Development Programme (UNDP 2001).

Chapter 3

1. Author's calculations for this and following statistics are based on data from Population Reference Bureau (2001).

2. Information on growth in access rates among the disabled is not available because the disabled were included as a separate category in the NTIA research for the first time in 2000.

3. The classic S-shaped curve of innovation diffusion is marked by a process in which a small number of *innovators* and *early adopters* first take up an innovation (the bottom curve of the S), followed by the large bulk of *early majority* and *late majority* (as the S curve shoots up), leaving the *laggards* to get on board over time (the top curve of the S); see Rogers 1962. For example, with Internet access, the innovators and early adopters came online from 1970 to 1995, the early and late majorities will have likely come by 2005, and the so-called laggards may take several more decades after that.

4. In several of these countries the Purchasing Power Parity (PPP) GNP is actually much higher than the US$ GDP, reflecting the fact that some goods (such as food and rent) are actually much cheaper than world market prices when purchased locally. For example, China's PPP GNP is $3,291, as compared with that country's US$ GDP of $450. However, the world price of computers does not vary much (and, in fact, tends to be higher in developing countries because of import expenses and tariffs), so it is actually the international currency capita that matters in this case, not the PPP GNP.

5. Information on the people's computer comes from the laboratory team's Web site, <http://www.luar.dcc.ufmg.br>, and from personal interviews with laboratory team member Wagner Meira, Jr., in August 2001.

6. Information on the Simputer reported in this section is from <http://www.simputer.org> and from personal interviews with Swami Manohar of the Simputer Trust in Bangalore, India, July 2001.

7. Posted on May 2, 2001; downloaded May 20, 2001, from <http://slashdot.org/articles/01/05/02/1822219.shtml>.

8. Posted on May 2, 2001; downloaded May 20, 2001, from <http://slashdot.org/articles/01/05/02/1822219.shtml>.

9. This would represent a deregulation of the "last mile," but the remainder of the telecommunications system would still be regulated.

10. Personal communication, July 2001, A. Jhunjhunwala.

Chapter 4

1. See, for example, "Web Surpasses One Billion Documents" (2000).

2. Data taken from the Internet Software Consortium, <http://www.isc.org/ds/host-count-history.html>.

3. See <http://www.healthnet.org>.

4. For further information, see <http://www.contentbank.org>.

5. See the Web Accessibility Initiative Web site, http://www.w3.org/wai/. Among other things, the site includes a useful overview of how disabled people use the Web (World Wide Web Consortium 2001b), a prioritized checklist for Web designers (World Wide Web Consortium 2001a), and a technical overview of user guidelines (World Wide Web Consortium 2001c). Also see Nielsen (1999).

6. Information taken from the Camfield Estates Web site, <http://www.camfieldestates.net/>, and from an interview in June 2001 with project organizers Randall Pinkett and Jeffrey Robinson.

7. Information taken from the HarlemLive Web site, <http://www.harlemlive.org/>, and from interviews with Rahsaan Harris in June and August 2001.

8. See <http://www.olelo.hawaii.edu/OP/resources/leoki.html>.

Chapter 5

1. A project organizer in India expressed this sentiment to me in a particularly poetic way. In emphasizing the importance of developing users' skills and knowledge, he said, "A computer is like a knife; a surgeon can wield a knife to help cure people of disease, while a lot of others can just use a knife to cut up mango." (Interview with Naveen Prakash, Gyandoot Project Manager, Dhar, India, July 2001).

2. The earliest American newspaper, magazine, or journal reference that I found via a search on the Lexis Nexis database was a 1981 article in the *Washington Post* (Milloy 1981).

3. In this context, I am referring to critical literacy in its most general form, i.e., the ability to read critically. The term *critical literacy* is also used in a more neo-Marxian framework, implying literacy that recognizes unequal power distribution. See discussion in Lankshear (1994).

4. Information on CDI is from <http://www.cdi.org.br/>, from internal documents provided by CDI, from visits to CDI schools, and from interviews with CDI representatives in Brazil in August 2001.

5. Information on Playing2Win is from the organization's Web site, from presentations by Playing2Win Director Rahsaan Harris at the national CTCNet conference in San Diego, California, June 15–17, 2001, and from a personal interview with Harris conducted in San Diego in June 2001.

6. Information on this project and school comes from <http://equity4.clmer.csulb.edu/netshare/cti/%20for%20psrtec%20website/Amada%20and%20Michelle/>, from an interview with teachers Michelle Singer and Amada Irma H. Perez (May 2001), and from an interview with Kevin Rocap of the Center for Language Minority Education and Research in Long Beach, California (May 2001).

7. Information on Foshay is from <http://www.foshay.k12.ca.us/>, personal observations at the school (May 2001), and interviews with school administrators and teachers (May 2001).

8. Information on educational technology in Egypt is taken from my own three-year ethnographic research there from 1998 to 2001, which included visits to more than twenty schools, meetings with hundreds of educators, access to documents from governmentaled and nongovernmental agencies, participation in technology training sessions of Egyptian educators, and involvement in electronic discussion groups related to educational technology in Egypt.

9. For critiques of the Egyptian educational system, see Birdsall and Londoño (1997); Fergany, Farmaz, and Wissa (1996); Fergany (1998); Jarrar and Massialas (1992); Tawila et al. (2000); and Sarhaddi Nelson (2001).

10. Information about educational technology in China is based on my research trips to Hong Kong (1999 and 2001), Suzhou (1999), Nanjing (2001), and Beijing (2001); my discussions with faculty members from Beijing Normal University, Capital Normal University, and Beijing University in April and August 2001; and documents provided by project coordinators at Beijing Normal University.

Chapter 6

1. Personal communication, June 2001, Randall Pinkett.

2. Personal communication, June 2001, Ricardo Gutierrez.

3. Whether use of the Internet causes social isolation is a point that has been hotly debated by sociologists. Another large-scale study found that heavy users of the Internet suffered depression, presumably because Internet use took time and energy away from face-to-face interactions with friends and kin (Kraut, Patterson et al. 1998). However, a follow-up study led by the same researcher, using more up-to-date data, reversed the original view (Kraut et al. 2002). For further discussion of this issue, see the November 2001 (vol. 45, no. 3) issue of the *American Behavioral Scientist*, a special issue of ten papers devoted to the theme of the Internet in everyday life.

4. Sociologists have a range of views as to the different levels of social capital. My own three-level (micro-, macro, meso-) approach draws on a number of perspectives, including those of Woolcock (1998), Krishna (2000), Turner (2000), and Wellman et al. (2001).

5. Interview with S. Regunathan, Principal Secretary for Information Technology, Government of New Delhi, July 2001.

6. Information on the Community Digital Initiative comes from interviews with project director Richard Chabran in April and June 2001, a visit to the center in June 2001, and the Web sites of the project (<http://cdi.ucr.edu/>) and its umbrella group, Computers in Our Future (<http://www.ciof.org/>).

7. Personal communication, July 2001, K. G. Rajamohan.

8. Information on the Bresee Community Center and Cyberhood comes from a visit to the center and interviews with Bresee staff in April 2001 and from the Bresee Foundation's Web site <http://www.bresee.org/>.

9. See <http://www.bridges.com/>. There is also an online version called eChoices, <http://www.echoices.com/>.

10. Interview with Jeff Carr, Director of Bresee Community Center, April 2001.

11. Information on M. S. Swaminathan Research Foundation projects comes from a visit to their headquarters and rural projects in July 2001 and interviews with members of their staff. Further information is available from <http://www.mssrf.org/>.

12. Information on ISIS is from <http://www2.gribus.at/> and from European Commission (2001a).

13. The European Computer Driving License is a Europe-wide qualification that enables people to demonstrate their competence in computer skills; see <http://www.ecdl.com/>.

14. Information on the Kothmale project is from <http://www.kothmale.net/> and from a report by A. Gumucio Dagron (2001, 127–132).

15. My information about land record systems in India, and about the computerized land record system of Karnataka, comes from the following sources: personal interview in July 2001 with Rajeef Chawla, Additional Secretary of the Revenue Department of the Government of Karnataka; Karnataka government documents on the system provided by Mr. Chawla in July 2001; a visit to a land record office in Bangalore, India, and interviews with the staff and clients in July 2001; interviews with small farmers and development workers in Pondicherry, Tamil Nadu, and Madya Pradesh in July 2001; and an interview with Bhawti Soleki, Assistant Vice President for Social Development of the Infrastructure Development Finance Company in New Delhi, India, in July 2001.

16. The first experience with online voting in the United States was in the 2000 Arizona Democratic Primary election. An analysis of the voting patterns showed that the nonwhite, the unemployed, and the elderly were all significantly less likely to participate in Internet voting than the public at large and that this limited their proportion of the vote compared to prior elections (Alvarez and Nagler 2001). The authors of the study conclude that "if Internet voting were widely used in American politics, it would change the character of political representation, with some specific groups behind the digital divide (minorities, the unemployed, and the elderly) losing further political power" (1148). Stephen Pershing (2001), an attorney in the U.S. Department of Justice, Civil Rights Division, has similarly concluded that "Internet voting without protections for equal access

may violate section 2 of the Voting Rights Act" (1209), which prohibits denial or abridgement of voting rights due to race.

17. Information about Dhar and the Gyandoot project comes from the following sources: interviews with Naveen Prakesh, Gyandoot Project Manager, in July 2001 (and subsequent e-mail correspondence in August–September 2001); an interview with Rajesh Rajora, then Dhar District Magistrate, in July 2001; a recent book by Rajora (2002); visits to five Gyandoot project information kiosks and interviews with managers and users in July 2001; and electronic and paper documents provided by the Gyandoot project in July 2001.

18. There were 26,560 groups listed on October 6, 2001; see <http://dir.groups.yahoo.com/dir/Health_Wellness/>.

19. For example, 208 such groups were listed in Yahoo! Groups on December 11, 2001; see <http://dir.groups.yahoo.com/dir/Health_Wellness/Reproductive/Infertility/>.

20. See <http://indymedia.org/>.

21. See note 16 above.

Chapter 7

1. See, for example, Ellen Seiter's (1993; 2000) work on the social context of television and Claude Fischer's (1992) work on the telephone and the automobile in the United States.

2. Though the point Levinson makes is an important one, it was perhaps unwise of him to use the word *determinist* for both concepts, since the term is usually associated with the "hard" variant.

3. The initial founders of this research project still direct two of the leading research institutes on technology and society today. Kenneth Kraemer and James Danziger head up the Center for Research on Information Technology and Organizations (CRITO—<http://www.crito.uci.edu>) at University of California, Irvine. Rob Kling directs the Center for Social Informatics at Indiana University (<http://www.slis.indiana.edu/CSI/>). Kling also edits the journal *The Information Society*, which has been a flagship of this research tradition since its founding in 1981.

References

Abamedia. 1999. Propaganda in the propaganda state. <http://www.pbs.org/redfiles/prop/inv/prop_inv_ins.htm>. Retrieved May 10, 2001.

Accenture, Markle Foundation, and United Nations Development Programme. 2001. Creating a development dynamic: Final report of the digital opportunity initiative. <http://www.opt-init.org/framework.html>. Retrieved January 1, 2002.

Adelman, M. B., M. R. Parks, and T. L. Albrecht. 1987. Beyond close relationships: Support in weak ties. In *Communicating social support*, ed. T. L. Albrecht and M. B. Adelman, 126–147. Newbury Park, Calif.: Sage.

Agre, P. E. 1997. Building community networks. In *Reinventing technology, rediscovering community: Critical explorations of computing as a social practice*, ed. P. E. Agre and D. Schuler, 241–248. Greenwich, Conn.: Ablex.

———. 1998. Building an Internet culture. *Telematics and Informatics* 15 (3): 231–234.

———. 1999a. Growing a democratic culture: John Commons on the wiring of civil society. Paper presented at the Media in Transition Conference, Massachusetts Institute of Technology, Cambridge, Mass. <http://dlis.gseis.ucla.edu/people/pagre/commons.html>. Retrieved December 31, 2001.

———. 1999b. Information technology and democratic institutions. <http://gseis.ucla.edu/people/agre/ottawa.html>. Retrieved February 1, 2000.

———. 2001a. Institutions and the entrepreneurial self. <http://commons.somewhere.com/rre/2001/RRE.Institutions.and.the.html>. Retrieved December 20, 2001.

———. 2001b. Networking on the network. <http://dlis.gseis.ucla.edu/people/pagre/network.html>. Retrieved October 19, 2001.

Aichholzer, G., and R. Schmutzer. 2001. *The digital divide in Austria*. Vienna: Institute of Technology Assessment.

Alkalimat, A., and K. Williams. 2001. Social capital and cyberpower in the African American community: A case study of a community technology centre in the dual city. In *Community informatics: Shaping computer-mediated social networks*, ed. L. Keeble and B. Loader, 177–204. London: Routledge.

Alvarez, R. M., and J. Nagler. 2001. The likely consequences of Internet voting for political representation. *Loyola of Los Angeles Law Review* 34 (3): 1115–1153.

Amadeu da Silva, S. 2001. *Exclusão Digital: A miséria na era da informação.* São Paulo: Fundaçao Perseu Abramo.

ASCII. 2001. Jargon File 4.3.1. <http://www.tuxedo.org/~esr/jargon/html/entry/ASCII.html>. Retrieved January 2, 2002.

Askonas, P., and A. Stewart, eds. 2000. *Social inclusion: Possibilities and tensions*. Houndmills, England: Macmillan.

Barlow, J. P. 1996. Declaration of independence of cyberspace. <http://www.eff.org/~barlow/Declaration-Final.html>. Retrieved July 1, 1999.

Bartholomae, D. 1986. Inventing the university. *Journal of Basic Writing* 5 (1): 4–23.

Bateson, G. 1972. *Steps to an ecology of mind: A revolutionary approach to man's understanding of himself*. New York: Ballantine.

Báthory-Kitz, D. 1999. Web accessibility of the presidential candidate sites. <http://orbitaccess.com/presidential/>. Retrieved January 6, 2002.

Becker, H. J. 1993. Computer experience, patterns of computers use, and effectiveness—An inevitable sequence or divergent national cultures? *Studies in Educational Evaluation* 19 (Summer): 127–148.

———. 2000. Who's wired and who's not? *The Future of Children* 10 (2): 44–75.

Bell, D. 1973. *The coming of post-industrial society*. New York: Basic Books.

Binns, D. 2001. Review of de Soto's *The mystery of capital: Why capitalism triumphs in the West and fails everywhere else*. <http://www.fed.org/onlinemag/jan01/reviews.htm>. Retrieved February 21, 2002.

Birdsall, N., and O. C. Lesley. 1999. *Globalization, income distribution and education: Putting education to work in Egypt*. Cairo: Egyptian Center for Economic Studies.

Birdsall, N., and J. L. Londoño. 1997. *Inequality and human capital accumulation in Latin America (with some lessons for Egypt)*. Distinguished Lecture Series 7. Cairo: Egyptian Center for Economic Studies.

Blom, J.-P., and J. J. Gumperz. 1972. Social meaning in linguistic structures: Code-switching in Norway. In *Directions in sociolinguistics*, ed. J. J. Gumperz and D. Hymes, 407–434. New York: Holt, Rinehart and Winston.

Bloom, J. R. 1982. Social support, accommodation to stress and adjustment to breast cancer. *Social Science and Medicine* 16: 1329–1338.

Bolter, J. D. 1991. *Writing space: The computer, hypertext, and the history of writing*. Hillsdale, N.J.: Erlbaum.

———. 1996. Ekphrasis, virtual reality, and the future of writing. In *The future of the book*, ed. G. Nunberg, 253–272. Berkeley: University of California Press.

Bourdieu, P. 1986. The forms of capital. In *Handbook of theory and research for the sociology of education*, ed. J. G. Richardson, 241–258. Westport, Conn.: Greenwood Press.

Bourguignon, F., and C. Morrison. 1999. *The size distribution of income among world citizens*. Paris: Delta and University of Paris, Département et laboratoire d'économie théorique et appliquée.

Bowles, S., and H. Gintis. 1976. *Schooling in capitalist America: Educational reform and the contradictions of economic life*. New York: Basic Books.

Brown, D. C. 1980. *Electricity for rural America: The fight for the REA*. Westport, Conn.: Greenwood Press.

Brown, J. S., A. Collins, and P. Duguid. 1989. Situated cognition and the culture of learning. *Educational Researcher* 18 (1): 32–42.

Brown, J. S., and P. Duguid. 2000. *The social life of information*. Boston: Harvard Business School Press.

Bruner, J. S. 1972. *The relevance of education*, ed. Anita Gil. London: Allen and Unwin.

Buchner, B. J. 1988. Social control and the diffusion of modern telecommunications technologies: A cross-national study. *American Sociological Review* 53 (3): 446–453.

Burbules, N. C., and T. A. Callister, Jr. 2000. *Watch IT: The risks and promises of information technologies for education*. Boulder, Colo.: Westview Press.

Bush, V. 1945. As we may think. *Atlantic Monthly* 176: 101–108.

Buzato, M. E. K. 2001. O letramento eletrônico e o uso do computador no ensino de língua estrangeira: Contribuições para a foramação de professores. Master's thesis, Universadade Estadual de Campinas, Campinas, Brazil.

Byrne, D. 1999. *Social exclusion*. Buckingham, England: Open University Press.

Canfield, J., and M. V. Hansen. 1993. *Chicken soup for the soul*. Deerfield Beach, Fla.: Health Communications, Inc. (HCI).

Carter, T. F. 1925. *The invention of printing in China and its spread westward*. New York: Ronald Press, 1955.

Carvin, A. 2000. Mind the gap: The digital divide as the civil rights issue of the new millennium. <http://www.infotoday.com/MMSchools/Jan2000/carvin.htm>. Retrieved May 10, 2001.

———. 2001. Website language stats. <http://owa.benton.org/listserv/wa.exe?A2=ind0104andL=digitaldivideandD=1andT=0andO=DandF=landS=andP=11879>. Retrieved December 28, 2001.

Castells, M. 1993. The informational economy and the new international division of labor. In *The new global economy in the information age: Reflections on our changing world*, ed. M. Carnoy, M. Castells, S. S. Cohen, and F. H. Cardoso, 15–43. University Park: Pennsylvania State University Press.

———. 1997. *The power of identity*. Malden, Mass.: Blackwell.

———. 2000a. *End of millennium*. 2d ed. Malden, Mass.: Blackwell.

———. 2000b. *The rise of the network society*. 2d ed. Malden, Mass.: Blackwell.

———. 2001. *The Internet galaxy: Reflections on the Internet, business, and society*. New York: Oxford University Press.

Castells, M., and E. Kiselyova. 1995. *The collapse of Soviet communism: A view from the information society*. Berkeley: University of California Press.

Cattagni, A., and E. F. Westat. 2001. Internet access in U.S. public schools and classrooms: 1994–2000. National Center for Education Statistics. <http://nces.ed.gov/pubs2001/2001071.pdf>. Retrieved February 21, 2002.

Chambers, R. 1992. Rural appraisal: Rapid, relaxed and participatory. Discussion Paper 311. Brighton, U.K.: University of Sussex, Institute of Development Studies.

Charbonnier, G. 1973. "Primitive" and "civilized" peoples: A conversation with Claude Lévi-Strauss. In *The future of literacy*, ed. R. Disch. Englewood Cliffs, N.J.: Prentice-Hall.

Christensen, C. M. 1997. *The innovator's dilemma: When new technologies cause great firms to fail*. Boston: Harvard Business School Press.

Christensen, C. M., T. Craig, and S. Hart. 2001. The great disruption. *Foreign Affairs* 90 (2): 80–95.

Cisler, S. 2000. Subtract the digital divide. <http://www.mercurycenter.com/svtech/news/indepth/docs/soap011600.htm>. Retrieved December 28, 2001.

Cleaver, H. M. 1998. The Zapatista effect: The Internet and the rise of an alternative political fabric. *Journal of International Affairs* 51 (2): 621–640.

Clines, F. X. 2001. Wariness leads to motivation in Baltimore free-computer experiment. *New York Times,* May 24.

CNNIC (China Internet Network Information Center). 2000. *Seminannual survey report on the development of China's Internet 2000/7*. <http://www.cnnic.net.cn/develst/e-cnnic200007.shtml>. Retrieved December 20, 2001.

———. 2001. *Seminannual survey report on the development of China's Internet 2001/7*. <http://www.cnnic.net.cn/develst/rep200107-e.shtml>. Retrieved December 20, 2001.

Coleman, J. S. 1988. Social capital in the creation of human capital. *American Journal of Sociology* 94: S95–S120.

Colker, D. 2001. Stirring a virtual melting pot. *Los Angeles Times,* February 20, A1.

Collier, P. 1998. Social capital and poverty. Social Capital Initiative Working Paper 4. Washington, D.C.: World Bank.

Collins, A., J. S. Brown, and S. E. Newman. 1989. Cognitive apprenticeship: Teaching the crafts of reading, writing, and mathematics. In *Knowing, learning, and instruction,* ed. L. B. Resnick, 453–494. Hillsdale, N.J.: Erlbaum.

Collins, H., and D. Braga. 2001. Interação e interatividade em duas modalidades de ensino de leitura na Internet. Paper presented at the Brazilian Congress of Applied Linguistics, Belo Horizonte, Brazil.

Corea, S. 2000. Cultivating technological innovation for development. *Electronic Journal on Information Systems in Developing Countries* 2 (2): 1–15. <http://www.ejisdc.org/>. Retrieved December 20, 2001.

Crystal, D. 1997. *English as a global language.* Cambridge: Cambridge University Press.

Cuban, L. 1986. *Teachers and machines: The classroom use of technology since 1920.* New York: Teachers College Press.

———. 1993. *How teachers taught: Constancy and change in American classrooms, 1890–1980.* 2d ed. New York: Longman.

———. 2001. *Oversold and underused: Computers in classrooms, 1980–2000.* Cambridge, Mass.: Harvard University Press.

Cummins, J. 1984. *Bilingualism and special education: Issues in assessment and pedagogy.* Clevedon, England: Multilingual Matters.

Cummins, J., and D. Sayers. 1990. Education 2001: Learning networks and educational reform. *Computers in the Schools* 7 (1/2): 1–29.

———. 1995. *Brave new schools: Challenging cultural illiteracy through global learning networks.* New York: St. Martin's Press.

Cyberspeech. 1997. *Time* 149 (June 23): 23.

Danziger, J. N., W. H. Dutton, R. Kling, and K. L. Kramer. 1982. *Computers and politics: High technology in American local government.* New York: Columbia University Press.

Day, A., and M. Miller. 1990. Gabriel García Márquez on the misfortunes of Latin America, his friendship with Fidel Castro and his terror of the blank page. *Los Angeles Times Magazine,* September 2, 10.

de Castell, S., and A. Luke. 1986. Models of literacy in North American schools: Social and historical conditions and consequences. In *Literacy, society, and schooling,* ed. S. de Castell, A. Luke, and K. Egan, 87–109. New York: Cambridge University Press.

de Soto, H. 2000. *The mystery of capital: Why capitalism triumphs in the West and fails everywhere else.* New York: Basic Books.

De Vaney, A., S. Gance, and Y. Ma, eds. 2000. *Technology and resistance: Digital communications and new coalitions around the world.* New York: Peter Lang.

Dede, C. 1995. Testimony to the U.S. Congress, House of Representatives, Joint Hearing on Educational Technology in the 21st century. <http://www.virtual. gmu.edu/SS_research/cdpapers/congrpdf.htm>. Retrieved January 4, 2002.

———. 1997. Rethinking how to invest in technology. <http://www.ascd.org/ articles/9711el_dede.html>. Retrieved January 4, 2002.

The default language. 1999. *Economist* (May 15): 67.

Dell, M., and C. Fredman. 1999. *Direct from Dell: Strategies that revolutionized an industry.* New York: Harper Business.

Diamond, L. 1994. Rethinking civil society: Toward democratic consolidation. *Journal of Democracy* 5 (3): 5–17.

DiCarlo, L. 1997. Buying PCs directly means no muss, no fuss. *PC Week,* February 17, 18.

Dikhanov, Y., and M. P. Ward. 2000. Towards a better understanding of the global distribution of income. Washington, D.C.: World Bank.

DiMaggio, P. J., and E. Hargittai. 2001. From the "digital divide" to "digital inequality": Studying Internet use as penetration increases. Working Paper 19. Princeton, N.J.: Center for Arts and Cultural Policy Studies, Woodrow Wilson School, Princeton University.

DiMaggio, P. J., E. Hargittai, R. Neuman, and J. Robinson. 2001. Social implications of the Internet. *Annual Review of Sociology* 27: 307–336.

Dimaggio, P. J., and W. W. Powell. 1991. Introduction. In *The new institutionalism in organizational analysis,* ed. W. W. Powell and P. J. Dimaggio, 1–38. Chicago: University of Chicago Press.

Dimond, M. 1979. Social support and adaptation to chronic illness: The case of maintenance hemodialysis. *Research in Nursing and Health* 2: 101–108.

Dividing lines. 2001. *Education Week on the Web* 20 (May 10). <http://www.edweek.org/sreports/tc01/tc01article.cfm?slug=35divideintro.h20>. Retrieved February 22, 2002.

Eisenstein, E. L. 1979. *The printing press as an agent of change: Communications and cultural transformations in early-modern Europe.* Cambridge: Cambridge University Press.

European Commission. 2001a. e-Inclusion practices. Background Document 1 to the Working Document. <http://europa.eu.int/comm/employment_ social/soc-dial/info_soc/esdis/eincl_1practices.pdf>. Retrieved December 10, 2001.

———. 2001b. e-Inclusion: The information society's potential for social inclusion in Europe. Commission Staff Working Document SEC (2001)1428. <http://europa.eu.int/comm/employment_social/soc-dial/info_soc/esdis/ documents.htm>. Retrieved December 10, 2001.

Fairclough, N. 1989. *Language and power.* London: Longman.

FCC (Federal Communications Commission). 1999. FCC releases new telephone subscribership report. <http://www.fcc.gov/Bureaus/Common_Carrier/News_Releases/1999/nrcc9006.html>. Retrieved May 10, 2001.

Feenberg, A. 1991. *Critical theory of technology.* New York: Oxford University Press.

———. 1999a. No frills in the virtual classroom. *Academe* 85 (5). <http://www.aaup.org/SO99Feen.htm>. Retrieved November 20, 1999.

———. 1999b. *Questioning technology.* London: Routledge.

Feldman, A., C. Konold, and B. Coulter. 2000. *Network science, a decade later: The Internet and classroom learning.* Mahwah, N.J.: Erlbaum.

Fergany, N. 1998. *Human capital and economic performance in Egypt.* Mimeo. Cairo.

Fergany, N., I. Farmaz, and C. Wissa. 1996. *Enrollment in primary education and cognitive achievement in Egypt: Change and determinants.* Cairo: Almishkat Centre for Research and Training.

Fischer, C. S. 1992. *America calling: A social history of the telephone to 1940.* Berkeley: University of California Press.

Foster, W., and S. Goodman. 2000. The diffusion of the Internet in China. Stanford University Center for International Security and Cooperation. <http://cisac.stanford.edu/docs/chinainternet.pdf>. Retrieved May 5, 2001.

Freire, P. 1970. The adult literacy process as cultural action for freedom. *Harvard Educational Review* 40: 205–212.

———. 1994. *Pedagogy of the oppressed.* 3d ed. New York: Continuum.

Freire, P., and D. Macedo. 1987. *Reading the word and the world.* Hadley, Mass.: Bergin and Garvey.

Gee, J. P. 1996. *Social linguistics and literacies.* London: Taylor and Francis.

Gee, J. P., G. Hull, and C. Lankshear. 1996. *The new work order: Behind the language of new capitalism.* St. Leonards, Australia: Allen and Unwin.

Ginsparg, P. 2001. Creating a global knowledge network. Paper presented at the Second Joint ICSU Press–UNESCO Conference on Electronic Publishing in Science. <http://arXiv.org/blurb/pg01unesco.html>. Retrieved January 4, 2001.

Gómez, R., P. Hunt, and E. Lamoureux. 1999. Telecentros y desarrollo social. *Chasqui: Revisa Latinoamericana de Comunicación* 66: 54–58.

Goodin, R. E. 1996. Institutions and their designs. In *The theory of institutional design,* ed. R. E. Goodin, 1–53. Cambridge: Cambridge University Press.

Goody, J., ed. 1968. *Literacy in traditional societies.* Cambridge: Cambridge University Press.

Goody, J., and I. Watt. 1963. The consequences of literacy. *Comparative Studies in History and Society* 5: 304–345.

Governo Electronico. 2001. Relátorio final—Oficina para inclusá social. <http://www.governoeletronico.gov.br/arquivos/inclusao_digital_relatorio_final. pdf>. Retrieved December 15, 2001.

Graddol, D. 1997. *The future of English*. London: British Council.

———. 1999. The decline of the native speaker. In *English in a changing world*, ed. D. Graddol and U. H. Meinhof, 57–68. Guildford, U.K.: Biddles.

Graham, S., J. Cornford, and M. Simon. 1996. The socio-economic benefits of a universal telephone network: A demand-side view of universal service. *Telecommunications Policy* 20: 1.

Granovetter, M. 1973. Strength of weak ties. *American Journal of Sociology* 8: 1360–1380.

Green, H., and L. Himelstein. 1998. A cyber revolt in health care. *Business Week*, October 19, 154.

Greenfield, P. M. 1972. Oral and written language: The consequences for cognitive development in Africa, the United States, and England. *Language and Speech* 15: 169–178.

Gumucio Dagron, A. 2001. *Making waves: Stories of participatory communication for social change*. New York: Rockefeller Foundation.

Gurstein, M. 2000. *Community informatics: Enabling communities with information and communications technologies*. Hershey, Pa.: Idea Group.

Haeri, N. 1997. The reproduction of symbolic capital: Language, state, and class in Egypt. *Current Anthropology* 38 (1): 795–805.

Hafner, K., and M. Lyon. 1996. *Where wizards stay up late: The origins of the Internet*. New York: Simon and Schuster.

Halliday, M. A. K. 1993. Towards a language-based theory of learning. *Linguistics and Education* 5 (2): 93–116.

Hampton, K. N. 2000. Examining community in the digital neighborhood: Early results from Canada's wired suburb. In *Digital cities: Technologies, experiences, and future perspectives*, ed. T. Ishida and K. Isbister, 194–208. Heidelberg, Germany: Springer-Verlag.

———. 2001a. Living the wired life in the wired suburb: Netville, glocalization and civil society. Ph.D. diss., University of Toronto.

———. 2001b. Broadband neighborhoods—connected communities. In *CHI 2001 extended abstracts*, ed. J. Jacko and A. Sears, 301–302. New York: Association for Computer Machinery.

Hampton, K. N., and B. Wellman. 1999. Netville on-line and off-line: Observing and surveying a wired suburb. *American Behavioral Scientist* 43 (3): 475–492.

———. 2001. Long distance community in the network society: Contact and support beyond Netville. *American Behavioral Scientist* 45 (3): 477–496.

Hanson, E. 2001. Globalization, inequality, and the Internet in India. Paper presented at the annual meeting of the International Studies Association, Chicago.

Hargittai, E. 1999. Weaving the Western Web: Explaining differences in Internet connectivity among OECD countries. *Telecommunications Policy* 23 (10/11): 701–718.

———. 2002a. How wide a Web: Inequalities in access to information in the age of the Internet. Ph.D. diss., Princeton University.

———. in press. Beyond logs and surveys: In-depth measures of people's Web use skills. *Journal of the American Society for Information Science and Technology Perspectives.*

Harnad, S. 1991. Post-Gutenberg galaxy: The fourth revolution in the means of production and knowledge. *Public-Access Computer Systems Review* 2 (1): 39–53.

He, K., and J. Wu. 2001. Innovative research to achieve the objectives of eight-year-old Chinese children's abilities to read and write: The experimenation of integrating information technology into language literacy education. Unpublished manuscript, Beijing Normal University, China.

Heeks, R. 1999. ICTs, poverty and development. Working Paper 5. Institute for Development Policy and Management, University of Manchester. <http://idpm.man.ac.uk/idpm/diwpf5.htm>. Retrieved August 20, 2001.

Heller, M. 1982. Language, ethnicity and politics in Quebec. Ph.D. diss., University of California at Berkeley.

Henriquez, J., W. Hollway, C. Urwin, C. Venn, and V. Walkerdine. 1984. *Changing the subject.* New York: Methuen.

Heydenrych, J. 2001. Computer mediated communication and WWW: Delivery modes and implementation variables—the case of the University of South Africa. <http://www.techknowlogia.org/>. Retrieved January 2, 2002.

Hirsch, E. D. 1987. *Cultural literacy: What every American needs to know.* Boston: Houghton Mifflin.

Hoffman, D. L., and T. P. Novak. 2001. The growing digital divide: Implications for an open research agenda. In *Understanding the digital economy: Data, tools, and research,* ed. E. Brynjolfsson and B. Kahin, 245–260. Cambridge, Mass.: MIT Press.

Hornberger, N. 1997. Language policy, language education, and language rights: Indigenous, immigrant, and international perspectives. Paper presented at the annual conference of the American Association for Applied Linguistics, Orlando, Florida.

How many online? 2001. <http://www.nua.com/surveys/how_many_online/>. Retrieved December 15, 2001.

Hunt, P. 2001. True stories: Telecentres in Latin America and the Caribbean. *Electronic Journal on Information Systems in Developing Countries* 4 (5): 1–17. <http://www.ejisdc.org/>. Retrieved February 22, 2002.

ICTs in rural poverty alleviation. 2001. *Economic and Political Weekly* 36 (March 17).

India. 2001. SIL International. <http://www.sil.org/ethnologue/countries/India. html>. Retrieved Sept. 1, 2001.

Indian languages. 2001. <http://indiansaga.com/languages/language_home. html>. Retrieved September 1, 2001.

Information Technology in Egypt. 1998. Cairo: American Chamber of Commerce in Egypt.

Jarboe, K. P. 2001. Inclusion in the information age: Reframing the debate. Athena Alliance. <http://www.athenaalliance.org/inclusion.html>. Retrieved December 15, 2001.

Jarrar, S. A., and B. G. Massialas. 1992. Arab Republic of Egypt. In *International handbook of educational reform,* ed. J. Cookson, W. Peter, A. R. Sadovnik, and S. F. Semel, 149–167. Westport, Conn.: Greenwood Press.

Jennings, T. 2001. ASCII: American standard code for information infiltration. World Power Systems. <http://www.wps.com/texts/codes/index.html>. Retrieved January 2, 2002.

Jhunjhunwala, A. 2000. Unleashing telecom and Internet in India. Paper presented at the Asia/Pacific Research Center, Stanford University. <http://www. tenet.res.in/Papers/unleash.html>. Retrieved December 30, 2001.

Kalathi, S., and T. C. Boas. 2001. The Internet and state control in authoritarian regimes: China, Cuba, and the counterrevolution. *First Monday* 6 (8). <http://www.firstmonday.dk/issues/issue6_8/kalathil/index.html>. Retrieved October 6, 2001.

Kaplan, N. 1995. E-literacies. *Computer-Mediated Communication Magazine* 2 (3): 3–35. <http://sunsite.unc.edu/cmc/mag/1995/mar/kaplan.html>.

Kelly, K. 1997. New rules for the new economy. *Wired* 5: 140–144, 186–194.

Klein, H. K. 1999. Tocqueville in cyberspace: Using the Internet for citizen's associations. *Information Society* 25: 213–220.

Kling, R. 1991. Computerization and Social Transformations. *Science, Technology, and Human Values* 16 (3): 342–367.

———. 1999. What is social informatics and why does it matter? *D-Lib Magazine* 5 (1). <http://www.dlib.org/dlib/january99/kling/01kling.html>. Retrieved December 15, 2001.

———. 2000. Learning about information technologies and social change: The contribution of social informatics. *Information Society* 16 (3): 1–36.

Kling, R., J. Fortuna, and A. King. 2001. The real stakes of virtual publishing: The transformation of E-Biomed into PubMed Central. CSI Working Paper 01–03. Indiana University Center for Social Informatics. <http://www.slis.indiana.edu/csi/wp/wp01-03B.html>. Retrieved January 5, 2002.

Kling, R., and W. Scacchi. 1979. Recurrent dilemmas of computer use in complex organizations. *Proceedings of the National Computer Conference* 48: 107–116.

————. 1982. The web of computing: Computer technology as social organization. In *Advances in Computers*, vol. 21, ed. M. C. Yovits, 3–85. New York: Academic Press.

Kraemer, K. L., J. Dedrick, and S. Yamashiro. 2000. Refining and extending the business model with information technology: Dell computer corporation. *Information Society* 16 (1): 5–21.

Kramsch, C., F. A'Ness, and E. Lam. 2000. Authenticity and authorship in the computer-mediated acquisition of L2 literacy. *Language Learning and Technology* 4 (2): 78–104.

Kranzberg, M. 1985. The information age: Evolution or revolution? In *Information technologies and social transformation*, ed. B. R. Guile, 35–54. Washington, D.C.: National Academy of Engineering.

Krashen, S. 1989. We acquire vocabulary and spelling by reading: additional evidence for the input hypothesis. *Modern Language Journal* 73: 440–464.

Kraut, R., S. Kiesler, B. Boneva, J. Cummings, V. Helgeson, and A. Crawford. 2002. Internet paradox revisited. *Journal of Social Issues* 58: 49–74.

Kraut, R., M. Patterson, V. Lundmark, S. Kiesler, T. Mukophadhyay, and W. Scherlis. 1998. Internet paradox: A social technology that reduces social involvement and psychological well-being. *American Psychologist* 53 (9): 1017–1031.

Kress, G. 1998. Visual and verbal modes of representation in electronically mediated communication: The potentials of new forms of text. In *Page to screen: Taking literacy into the electronic era*, ed. I. Snyder, 53–79. London: Routledge.

Kress, G., and T. van Leeuwen. 1996. *Reading images: The grammar of visual design*. London: Routledge.

Krishna, A. 2000. Creating and harnessing social capital. In *Social capital: A multifaceted perspective*, ed. P. Dasgupta and I. Serageldin, 71–93. Washington, D.C.

Kualono. 2001. *Leokī: Kikowaena Kelaka'a'ike 'Ōlelo Hawai'i*. <http://www.olelo.hawaii.edu/eng/information/leoki.html>. Retrieved January 1, 2002.

Labaton, S. 2001. New F.C.C. chief would curb agency reach. *New York Times*, February 7.

Lamberg, L. 1997. Computers enter mainstream psychiatry. *Journal of the American Medical Association* 278: 799–801.

Languages in Europe. 2002. <http://europa.eu.int/comm/education/languages/lang/europeanlanguages.html>. Retrieved January 1, 2002.

Lanham, R. A. 1993. *The electronic word: Democracy, technology, and the arts*. Chicago: University of Chicago Press.

Lankshear, C. 1994. *Critical literacy*. Belconnen, Australia: Australian Curriculum Studies Association.

Lave, J. 1988. *Cognition in practice: Mind, mathematics and culture in everyday life.* Cambridge: Cambridge University Press.

Lave, J., and E. Wenger, eds. 1991. *Situated learning: Legitimate peripheral participation.* Cambridge: Cambridge University Press.

Lazarus, W., and F. Mora. 2000. *Online content for low-income and underserved Americans: The digital divide's new frontier.* Santa Monica, Calif.: Children's Partnership.

Lerner, D. 1958. *The passing of traditional society.* New York: Free Press.

Lessig, L. 1999. *Code and other laws of cyberspace.* New York: Basic Books.

Levine, P. 2001. The Internet and civil society: Dangers and opportunities. <http://www.cisp.org/imp/may_2001/05_01levine.htm>. Retrieved October 6, 2001.

Levinson, P. 1997. *The soft edge: A natural history and future of the information revolution.* London: Routledge.

Lievrouw, L. A. 2000. The information environment and universal service. *Information Society* 16: 155–159.

Lin, N. 2001. *Social capital: A theory of social structure and action.* Cambridge: Cambridge University Press.

Littlewood, P., I. Glorieux, S. Herkommer, and I. Jonsson, eds. 1999. *Social exclusion in Europe: Problems and paradigms.* Aldershot, England: Ashgate.

Loader, B., B. Hague, L. Keeble, and D. Eagle, eds. 2001. *Community informatics: Shaping computer-mediated social networks.* London: Routledge.

Luria, A. R. 1976. *Cognitive development: Its cultural and social foundations.* Cambridge, Mass.: Harvard University Press.

Magretta, J. 1998. The power of virtual integration: An interview with Dell Computer's Michael Dell. *Harvard Business Review* (March–April): 73–84.

McConnell, S. 2000. A champion in our midst: Lessons learned from the impact of NGOs' use of the Internet. *Electronic Journal on Information Systems in Developing Countries* 2 (5): 1–14. <http://www.ejisdc.org/>. Retrieved December 20, 2001.

McLuhan, M. 1962. *The Gutenberg galaxy: The making of typographic man.* Toronto: University of Toronto Press.

Means, B., ed. 1994. *Technology and education reform: The reality behind the promise.* San Francisco: Jossey-Bass.

Milanovic, B. 1999. True world income distribution, 1988 and 1993: First calculation based on household surveys alone. Washington, D.C.: World Bank.

Milloy, C. 1981. D.C. schools: How to make do with less. *Washington Post,* November 15, B1.

Mitra, S. 1999. Minimally invasive education for mass computer literacy. *CSI Communications* (June): 12–16. Computer Society of India. <http://www.csi-india.org/csicomm.html>. Retrieved February 22, 2002.

Mokyr, J. 1990. *The lever of riches: Technological creativity and economic progress.* New York: Oxford University Press.

Morton, F. S., F. Zettelmeyer, and J. Silva-Risso. 2001. Consumer information and price discrimination: Does the Internet affect the pricing of new cars to women and minorities? Working Paper 8668. National Bureau of Economic Research. <http://papers.ssrn.com/sol3/papers.cfm?abstract_id=294106>. Retrieved January 4, 2002.

Mukherjee, N. 1993. *Participatory rural appraisal: Methodology and applications.* New Delhi: Concept Publishing.

Murray, D. E. 1995. *Knowledge machines: Language and information in a technological society.* London: Longman.

Neumann, P., and C. Uhlenküken. 2001. Assistive technology and the barrier-free city: A case study from Germany. *Urban Studies:* 38 (2): 367–376.

Nie, N. H., and L. Erbring. 2000. *Internet and society: A preliminary report.* Stanford: Stanford Institute for the Quantitative Study of Society. <http://www.stanford.edu/group/siqss/Press_Release/Preliminary_Report.pdf>. Retrieved December 20, 2001.

Nielsen, J. 1999. Disabled accessibility: The pragmatic approach. <http://www.useit.com/alertbox/990613.html>. Retrieved January 6, 2002.

Noble, D. 1998a. Digital diploma mills. Part 1: The automation of higher education. <http://communication.ucsd.edu/dl/ddm1.html>. Retrieved May 30, 1999.

―――. 1998b. Digital diploma mills. Part 2: The coming battle over online instruction. <http://communication.ucsd.edu/dl/ddm2.html>. Retrieved May 30, 1999.

NTIA (National Telecommunications and Information Administration). 2000. *Falling through the net: Toward digital inclusion.* Washington, D.C.

Nye, D. 1990. *Electrifying America: A social meaning of new technology, 1880–1940.* Cambridge, Mass.: MIT Press.

Ochs, E., and B. Shieffelin. 1984. Language acquisition and socialization: Three developmental stories and their implications. In *Culture theory: Essays on mind, self, and emotion,* ed. R. Shweder and R. Levine, 276–320. Cambridge: Cambridge University Press.

OECD (Organization for Economic Cooperation and Development). 2000. *OECD information technology outlook 2000: ICTs, e-commerce, and the information economy.* <http://www.oecd.org/>.

―――. 2001. *Communications outlook 2001.* <http://www.oecd.org/>.

O'Leary, S. D. 2000. Falun Gong and the Internet. *Online Journalism Review,* June 15. <http://ojr.usc.edu/content/story.cfm?request=390>. Retrieved January 1, 2002.

Olson, D. R. 1977. From utterance to text: The bias of language in speech and writing. *Harvard Educational Review* 47 (3): 257–281.

————. 1994. *The world on paper*. Cambridge: Cambridge University Press.

Ong, W. 1982. *Orality and literacy: The technologizing of the word*. London: Routledge.

Oppenheimer, T. 1997. The computer delusion. *Atlantic Monthly* 289 (July): 48–62.

Orr, J. 1966. *Talking about machines: An ethnography of a modern job*. Ithaca, N.Y.: IRL Press.

Osin, L. 1998. Computers in education in developing countries: Why and how? *Education and Technology Technical Notes Series* 13 (1). Washington, D.C.: World Bank.

Papert, S. 1980. *Mindstorms: Children, computers, and powerful ideas*. New York: Basic Books.

Pastore, M. 2000. Web pages by language. <http://cyberatlas.internet.com/big_picture/demographics/article/0,1323,5901_408521,00.html>.   Retrieved August 20, 2001.

Patterson, R., and E. J. Wilson. 2000. New IT and social inequality: Resetting the research and policy agenda. *Information Society*: 16 (1): 77–86.

PCs and teachers omitted from new computer science curriculum. 2000. *Egyptian Gazette,* September 22, 2.

Perez, C., and L. Soete. 1988. Catching up in technology: Entry barriers and windows of opportunity. In *Technical change and economic theory*, ed. G. Dossi, C. Freeman, R. Nelson, G. Silverberg, and L. Soete, 458–479. London: Pinter.

Pershing, S. B. 2001. The voting rights act in the Internet age: An equal access theory for interesting times. *Loyola of Los Angeles Law Review* 34 (3): 1171–1211.

Piaget, J. 1970. *Science of education and the psychology of the child*. New York: Orion Press.

Pomfret, J. 2000. A low-key revolution: China's gays are coming out of the closet. *International Herald Tribune*, January 25, 2001.

Population Reference Bureau. 2001. *2001 world population datasheet.* <http://www.prb.org/>. Retrieved December 15, 2001.

Poster, M. 1997. Cyberdemocracy: Internet and the public sphere. In *Internet culture*, ed. D. Porter, 201–217. New York: Routledge.

Postman, N. 1993. *Technopoly: The surrender of culture to technology*. New York: Vintage Books.

Potashnik, M. 1996. Chile's learning network. *Education and Technology Technical Notes Series* 1 (2). Washington, D.C.: World Bank.

Preston, P., and R. Flynn. 2000. Rethinking universal service: Citizenship, consumption norms, and the telephone. *Information Society*: 16: 91–98.

Prinsloo, M., and M. Breir, eds. 1996. *The social uses of literacy: Theory and practice in contemporary South Africa*. Philadelphia: Sached Books.

Proenza, F., J. Bastidas-Buch, and G. Montero. 2001. Telecenters for socio-economic and rural development in Latin America and the Caribbean. Inter-American Development Bank. <http://www.iadb.org/regions/itdev/telecenters>. Retrieved October 16, 2001.

Putnam, R. 1993. The prosperous community: Social capital and public life. *American Prospect* 13: 35–42.

———. 2000. *Bowling alone: The collapse and revival of American community*. New York: Simon and Schuster.

Rajora, R. 2002. *Bridging the digital divide: Gyanoot, the model for community networks*. New Delhi: Tata–McGraw Hill.

Redden, G. 2001. Networking dissent: The Internet and the anti-globalization movement. *Mots Pluriels* 18. <http://www.arts.uwa.edu.au/MotsPluriels/MP1801gr.html>. Retrieved October 6, 2001.

Reich, R. 1991. *The work of nations: Preparing ourselves for 21st century cap-italism*. New York: Knopf.

Renfrew, C. 1984. *Electricity, industry, and class in South Africa*. Albany: State University of New York Press.

Resnick, P. 2002. Beyond bowling together: Sociotechnical capital. In *Human computer interaction in the new millennium*, ed. J. Carroll, 247–272. New York: Addison-Wesley.

Rheingold, H. 2000. *The virtual community: Homesteading on the electronic frontier*. 2d ed. Cambridge, Mass.: MIT Press.

Robison, K. K., and E. M. Crenshaw. 2000. Cyber-space and postindustrial transformations: A cross-national analysis of Internet development. Paper presented at the annual meeting of the American Sociological Association, Washington, D.C. <http://home.columbus.rr.com/krisrobison/robisoncrenshaw cyber1ssrr.pdf>. Retrieved December 31, 2001.

Rodan, G. 1998. The Internet and political control in Singapore. *Political Science Quarterly* 113 (Spring).

Rogers, E. M. 1962. *Diffusion of innovations*. New York: Free Press.

Ronfeldt, D., and J. Arquilla. 2001. Networks, netwars, and the fight for the future. *First Monday* 6 (10). <http://www.firstmonday.org/issues/issue6_10/ronfeldt/>. Retrieved January 5, 2001.

Ronfeldt, D., J. Arquilla, G. E. Fuller, and M. Fuller. 1998. *The Zapatista social netwar in Mexico*. Santa Monica, Calif.: RAND Corp.

Rosenthal, E. 2001. China jails six for spreading sect's material. *New York Times*, December 24, A9.

Salamon, L. M. 1994. The rise of the nonprofit sector. *Foreign Affairs* 73 (4): 109–122.

Sandholtz, J. H., C. Ringstaff, and D. C. Dwyer. 1997. *Teaching with technology: Creating student-centered classrooms.* New York: Teachers College Press.

Sarhaddi Nelson, S. 2001. Egyptians' obsession with grades fails to nurture creative thinkers. *Los Angeles Times*, January 21, A8.

Schank, R. C., and C. Cleary. 1995. *Engines for education.* Hillsdale, N.J.: Erlbaum.

Schaub, M. 2000. English in the Arab Republic of Egypt. *World Englishes* 19 (2): 225–238.

Schecter, S., D. Sharken-Taboada, and R. Bayley. 1996. Bilingual by choice: Latino parents' rationales and strategies for raising children with two languages. *Bilingual Research Journal* 20 (2): 261–281.

Schement, J. R., and S. C. Forbes. 2000. Identifying temporary and permanent gaps in universal service. *Information Society* 16 (2): 117–126.

Schofield, J. W., and A. L. Davidson. in press. Achieving equality of student Internet access within schools. In *The social psychology of group identity and social conflict*, ed. A. H. Eagly, R. M. Baron, and V. L. Hamilton. Washington, D.C.: APA Books.

Scribner, S., and M. Cole. 1981. *The psychology of literacy.* Cambridge, Mass.: Harvard University Press.

SDRTA (San Diego Regional Technology Alliance). 2001. Mapping a future for digital connections: A study of the digital divide in San Diego County. <http://www.sdrta.org/sdrta/aboutsdrta/RTA_Report_0201.pdf>. Retrieved September 7, 2001.

Seiter, E. 1993. *Sold separately: children and parents in consumer culture.* New Brunswick, N.J.: Rutgers University Press.

———. 2000. *Television and new media audiences.* Oxford: Oxford University Press.

Selfe, C. 1990. Technology in the English classroom: Computers through the lens of feminist theory. In *Computers and community: Teaching composition in the twenty-first century*, ed. C. Handa, 118–139. Portsmouth, N.H.: Heinemann.

Selfe, C., and R. J. Selfe. 1994. The politics of the interface: Power and its exercise in electronic contact zones. *College Composition and Communication* 45 (4): 480–504.

Serageldin, I., and C. Grootaert. 2000. Defining social capital: An integrated view. In *Social capital: A mulifaceted perspective*, ed. P. Dasgupta and I. Serageldin, 40–58. Washington, D.C.: World Bank.

Shetzer, H., and M. Warschauer. 2000. An electronic literacy approach to network-based language teaching. In *Network-based language teaching: Concepts and practice*, ed. M. Warschauer and R. Kern, 171–185. New York: Cambridge University Press.

Singer, C., E. J. Holmyard, A. R. Hall, and T. I. Williams. 1958. *A history of technology: The industrial revolution, c. 1750 to c. 1850*. Vol. 4. Oxford: Clarendon Press.

Singh, J. P. 1999. *Leapfrogging development? The political economy of telecommunications restructuring*. Albany: State University of New York Press.

Smith, J. 2001. The mystery of capital. Review. *Geonomist* 9 (3). <http://www.progress.org/geonomy/geonom93.htm>. Retrieved October 2, 2001.

Smith, P. J., and E. Smythe. 1999. Globalization, citizenship and technology: The MAI meets the Internet. *Canadian Foreign Policy* 7 (2): 83–105.

Spyd3r. 1998. The history of hacking. Help Net Security. <http://www.net-security.org/text/articles/history.shtml>. Retrieved February 23, 2002.

Stanley, L. 2001. *Beyond access*. Occasional Paper 2. San Diego, Calif.: UCSD Civic Collaborative.

Stewart, A. 2000. Social inclusion: An introduction. In *Social inclusion: Possibilities and tensions*, ed. P. Askonas and A. Stewart, 1–16. Houndmills, England: Macmillan.

Street, B. 1984. *Literacy in theory and practice*. Cambridge: Cambridge University Press.

———. 1993. Introduction: The new literacy studies. In *Cross-cultural approaches to literacy*, ed. B. V. Street, 1–21. Cambridge: Cambridge University Press.

———. 1995. *Social literacies: Critical approaches to literacy in development, ethnography and education*. London: Longman.

Sunstein, C. 2001. *Republic.com*. Princeton, N.J.: Princeton University Press.

Tannen, D. 1994. *Gender and discourse*. New York: Oxford University Press.

Tauscher, L., and S. Greenberg. 1997. How people revisit Web pages: Empirical findings and implications for the design of history systems. *International Journal of Human Computer Studies* 47 (1): 97–138.

Tawila, S., C. B. Lloyd, B. S. Bensch, and H. Wassef. 2000. *The school environment in Egypt: A situational analysis of public preparatory schools*. Cairo: Population Council.

Theil, H. 1967. *Economics and information theory*. Chicago: Rand McNally.

Thierer, A. D. 2001. How free computers are filling the "digital divide": A PowerPoint presentation. Heritage Foundation. <http://www.heritage.org/features/powerpoint/digitaldivide/>. Retrieved February 23, 2002.

Tocqueville, A. de. 1835. *Democracy in America*. 2 vols. New York: Vintage Books, 1937, 1945.

Todd, L., and I. Hancock. 1987. *International English Usage*. New York: NYU Press.

Tukey, J. W. 1962. The future of data analysis. *Annals of Mathematical Statistics* 33: 13–14.

Tuman, M. 1992. *Word perfect: Literacy in the computer age*. Pittsburgh: University of Pittsburgh Press.

Turner, J. H. 2000. The formation of social capital. In *Social capital: A multi-faceted perspective*, ed. P. Dasgupta and I. Serageldin, 94–146. Washington, D.C.: World Bank.

Turner, J. W., J. A. Grube, and J. Meyers. 2001. Developing an optimal match within online communities: An exploration of CMC support communities and traditional support. *Journal of Communication* 51 (2): 231–251.

UNDP (United Nations Development Programme). 1999a. *China human development report: Transition and the state*. Oxford University Press (China). <http://www.unchina.org/undp/press/html/cnhdr99.htm>. Retrieved February 23, 2002.

———. 1999b. *Human development report 1999: Globalization with a human face*. <http://www.un.org/Pubs/textbook/e99hdp.htm>. Retrieved February 23, 2002.

———. 2000. *Human development report 2000: Human rights and human development*. <http://www.undp.org/hdr2000/home.html>. Retrieved February 23, 2002.

———. 2001. *Human development report 2001: Making new technologies work for human development*. <http://www.undp.org/hdro/>. Retrieved February 23, 2002.

United States Census Bureau. 1995. *Statistical abstract of the United States*. Washington, D.C.

Verdisco, A., and J. C. Navarro. 2000. Costa Rica: Teacher training for education technology. <http://www.techknowlogia.org/>. Retrieved December 31, 2001.

Vygotsky, L. S. 1978. *Mind in society*, ed. M. Cole, V. John-Steiner, S. Scribner, and E. Souberman. Cambridge, Mass.: Harvard University Press.

———. 1981. The genesis of higher mental functions. In *The concept of activity in Soviet psychology*, ed. J. V. Wertsch, 144–188. Armonk, N.Y.: M. E. Sharpe.

Wade, R. 2001. Winners and losers. *Economist*, April 28, 79–82.

Wales, J. A. 2001a. Equity and access to higher education for underrepresented students: Can advanced placement opportunities through online learning make a difference? The implications for technology. Unpublished manuscript, University of California at Irvine.

————. 2001b. Online AP macroeconomics. Unpublished manuscript, University of California at Irvine.

————. 2001c. Online introduction to computer science and the C programming language. Unpublished manuscript, University of California at Irvine.

Wallsten, S. J. 2001. An econometric analysis of telecom competition, privatization, and regulation in Africa and Latin America. *Journal of Industrial Economics* 49 (1): 1–20.

Walton, A. 1999. Technology vs. African-Americans. *Atlantic Monthly* 283 (1): 14–18.

Warschauer, M. 1997. Computer-mediated collaborative learning: Theory and practice. *Modern Language Journal* 81 (3): 470–481.

————. 1998. Technology and indigenous language revitalization: Analyzing the experience of Hawai'i. *Canadian Modern Language Review* 55 (1): 140–161.

————. 1999. *Electronic literacies: Language, culture, and power in online education*. Mahwah, N.J.: Erlbaum.

————. 2000a. The changing global economy and the future of English teaching. *TESOL Quarterly* 34: 511–535.

————. 2000b. Technology and school reform: A view from both sides of the tracks. *Education Policy Analysis Archives* 8 (4). <http://epaa.asu.edu/epaa/v8n4.html>. Retrieved January 2, 2002.

————. 2001a. The allures and illusions of modernity: Technology and educational reform in Egypt. Unpublished manuscript.

————. 2001b. Language choice online: Globalization vs. identity in the age of information. Unpublished manuscript.

————. 2001c. Singapore's dilemma: Control vs. autonomy in IT-led development. *Information Society* 17 (4): 305–311.

Warschauer, M., and K. Donaghy. 1997. Leokī: A powerful voice of Hawaiian language revitalization. *Computer Assisted Language Learning* 10 (4): 349–362.

Warschauer, M., G. Refaat, and A. Zohry. 2000. Language and literacy online: A study of Egyptian Internet users. Paper presented at the annual meeting of the American Association for Applied Linguistics, Vancouver, Canada.

Web surpasses one billion documents. 2000. <http://www.inktomi.com/new/press/2000/billion.html>. Retrieved August 1, 2001.

Weedon, C. 1987. *Feminist practice and poststructuralist theory*. London: Blackwell.

Wellman, B., A. Q. Haase, J. Witte, and K. Hampton. 2001. Does the Internet increase, decrease, or supplement social capital?: Social networks, participation, and community commitment. *American Behavioral Scientist* 45 (3): 437–456.

Wells, G., and G. L. Chang-Wells. 1992. *Constructing knowledge together*. Portsmouth, N.H.: Heinemann.

Wenglinsky, H. 1998. Does it compute? The relationship between educational technology and student achievement in mathematics. Policy Information Report. Educational Testing Service. <ftp://ftp.ets.org/pub/res/technolog.pdf>. Retrieved January 5, 2002.

West, C. 1990. Not just "doctor's orders": Directive-response sequences in patients' visits to women and men physicians. *Discourse and Society* 1 (1): 85–113.

Whalen, J., and E. Vinkhuyzen. 2000. Expert systems in (inter)action: Diagnosing document machine problems over the telephone. In *Workplace studies: Recovering work practice and informing system design*, ed. P. Luff, J. Hindmarsh, and H. Christiapp, 92–140. New York: Cambridge University Press.

Willis, P. E. 1977. *Learning to labor: How working class kids get working class jobs*. New York: Columbia University Press.

Wilson, E. 2000. *Briefing the President*. Internet Policy Institute. <http://www.internetpolicy.org/briefing/ErnestWilson0700.html>. Retrieved May 10, 2001.

Wilson, W. H. 1998. I ka 'olelo Hawai'i ke ola [Life is found in the Hawaiian language]. *International Journal of the Sociology of Language* 132: 123–137.

Winner, L. 1986. *The whale and the reactor: The search for limits in a technological society*. Chicago: University of Chicago Press.

Woolcock, M. 1998. Social capital and economic development: Toward a theoretical synthesis and policy framework. *Theory and Society* 27: 151–208.

World development report 1998/99: Knowledge and development. Washington, D.C.: World Bank.

World development report 2000/01: Attacking poverty. Washington, D.C.: World Bank.

World Wide Web Consortium. 2001a. Checklist of checkpoints for Web content accessibility guidelines 1.0. <http://www.w3.org/tr/wai-webcontent/full-checklist.html>. Retrieved January 6, 2002.

———. 2001b. How people with disabilities use the Web. <http://www.w3.org/wai/eo/drafts/pwd-use-web/overview.html>. Retrieved January 6, 2002.

———. 2001c. User agent accessibility guidelines 1.0. <http://www.w3.org/tr/uaag10/>. Retrieved January 6, 2002.

Wysocki, B. 1999. Dell or be delled. *Wall Street Journal*, May 10, A1.

Yeo, S., and A. Mahizhnan. 1999. Censorship: Rules of the game are changing. *Sunday Times*, August 15, 34–35.

Young, J. R. 2001. Does "digital divide" rhetoric do more harm than good? *Chronicle of Higher Education*, November 9. <http://chronicle.com/free/v48/i11/11a05101.htm>. Retrieved December 10, 2001.

Zanini, M., and S. J. A. Edwards. 2001. The networking of terror in the information age. In *Networks and netwars: The future of terror, crime, and militancy*, ed. J. Arquilla and D. Ronfeldt, 29–58. Santa Monica, Calif.: RAND Corp.

Zook, M. A. 2001a. Domain names worldwide. <http://www.zooknic.com/ Domains/international.html>. Retrieved December 28, 2001.

———. 2001b. Domains per capita. <http://www.zooknic.com/Domains/ Domains_per_capita.pdf>. Retrieved December 28, 2001.

———. 2001c. Old hierarchies or new networks of centrality? The global geography of the Internet content market. *American Behavioral Scientist* 44 (10): 1679–1696.

Zuboff, S. 1988. *In the age of the smart machine: The future of work and power.* New York: Basic Books.

Index

Academic institutions, 208–209
Access to information technology. *See also* Affordability; Content of digital resources; Electronic literacies; Rural areas
 comprising a broad array of factors, 46–48, 50, 51, 58, 199, 213
 contact with other users affecting, 156–157
 cost of computers and operating software, 32, 47, 62–63
 cost of Internet access, 52
 cultural barriers to, 45–46
 educational level and literacy affecting, 56, 57–58, 59
 lack of linguistic diversity, 92–96
 physical access as only one aspect of, 118–119
 race and ethnicity as a factor, 37, 55, 57–58
 socioeconomic factors, 29, 37, 49–52, 54–58, 59
Access models. *See* Conduits model of access; Devices model of access; Literacy model of access
Affordability
 cost of computers and operating software, 32, 47, 62–63
 cost of Internet access, 52
 plans to develop low-cost systems, 65–69
African Americans

Internet access rates for, 7, 55, 56, 57–58
 television diffusion rates for, 37
Africa, sub-Saharan, 19
Agency of International Development (USIA), 5
Agre, P. E., 187–188, 211, 212
Amazon.com, 64–65
American Standard Code for Information Exchange (ASCII), 203
Antiglobalism movement, 192, 193–195
Apprenticeship and mentoring, 121
Arabic language computing, 59, 101–102
ARPANET, 24
ASCII, 203
Asian Americans
 Internet access rates, 56
 school computer use by, 130–131
Associational technologies. *See* Information and communication technology (ICT); Networks
Audiovisual media, 27, 87
Automobile industry
 information-based capitalism influencing, 15–16

Bangladore, India, 60, 85
Bateson, Gregory, 110
Behavioral change, 211–212
Beijing, China, 61

Feenberg, A., 209–210
Female-headed households
  Internet access rates for, 57–58
Financial information online, 88
Finland, broad Internet access in, 52,
  53
Foshay Learning Center, Los Angeles,
  137–138
France
  low Internet connectivity in, 52, 53
  Minitel system in, 53, 62
Frankfurt school, 209
Fresa Project, 135–137

Gee, J. P., 39, 45
Gender
  Internet access and, 55, 61
  literacy and, 46
George Foundation, 85
Germany
  electrification in, 34
Global capitalism, 12–13, 15–18
Global English, 94–96
GNP. *See* Gross National Product
  (GNP)
Government
  citizen access to public documents
    and data, 88, 173–177, 183–184
  citizen access to resources of, 88,
    172, 173–174
  interactive citizen feedback
    programs, 172–173, 177, 179–181
  providing electrification and
    infrastructure, 34, 35, 53
  supporting access programs, 2–3, 5,
    9, 53, 65–66, 75, 76, 78–79
Granovetter, M., 155
Graphics. *See* Visual media on the
  Internet
Greece, low Internet connectivity in,
  52, 53
Gross National Product (GNP),
  219n.4
  crossnational comparisons, 18–20,
    23
  and Internet diffusion, 59

Guandong, China, 61
Gurstein, M., 162, 163
Gutenberg revolution, 39–40,
  204–205
Gyandoot rural technology project,
  85, 91, 171–172, 179–181

Hampton, Keith N., 158
Handheld computing devices, 66–69
Hard vs. soft media determinism,
  204–205
Hargittai, Ezster, 50–51, 53
Harlem
  HarlemLive Internet-based youth
    publication, 92
  Playing2Win, 127–129
Harnad, Stephen, 25, 26, 27
Hawai'i
  ethnographic research in, 47
  native language content development
    in, 103–107
Health-related information and
  networking, 28–29, 188–191
  in India's village knowledge centers,
    85–86
He, K., and J. Wu, 144
High-income groups. *See* Elite groups
Hindi language, 102–103
Hirsch, E. D., 119
Hispanics
  Internet access rates of, 55, 56,
    57–58
  school computer use by, 130–131
  television diffusion rates for, 37
Hole-in-the-Wall computer kiosks, 1,
  85, 91, 163, 179
Holland, 34
Hornberger, Nancy, 107
Host domains. *See* Domains on the
  Internet
Human resources, 47, 109
Hypertext, 26

ICT. *See* Information and
  communication technology (ICT)
Identity formation, 93–95, 122

DATE DUE